Analysis Zero

Anthony G. O'Farrell
MSc, PhD, MRIA
Professor Emeritus of Mathematics
Maynooth University

CLÓ LOIGHIC/ LOGIC PRESS
Kilcock, Co. Kildare

©Logic Press 2015

All rights are reserved including the specific rights to translate, reprint, recite, perform, broadcast or reproduce in any way or to store in data banks all or any part of this book, without prior permission from the publisher.

LOGIC PRESS
The Maws, Kilcock, Naas
Co. Kildare, W23 D92N, Ireland
Tel: +353-1-628-7343
e-mail: logicpress@gmail.com
www.logicpress.ie

ISBN: 978-1-326-41581-5
This is the first paperback edition. The hard-cover first edition has ISBN 978-1-326-41433-7.

First published 2015

Anthony G. O'Farrell is the sole author of this work.

Published by Logic Press
Printed and distributed by Lulu.com
3101 Hillsborough St., Raleigh, NC 27606-5436, USA

This book on the foundations of Analysis
is offered
for the greater honour and glory of
God
the Ground of my being
in loving memory of my parents
Patrick O'Farrell (1920–1997)
and
Sheila, née Curtis (1919–2006)
who called me into being
and grounded me whenever necessary.

Contents

1 Sets and Relations   **1**
   1.1 Overview 1
   1.2 Logic 2
   1.3 Sets 5
   1.4 Relations 15
   1.5 Functions 18
   1.6 Equivalence Relations 24
   1.7 Partial Orders 26

2 Natural Numbers   **30**
   2.1 The Peano Postulates 30
   2.2 Ordering $\mathbb{N}$ 33
   2.3 Inductive Definitions 39
   2.4 Binary Operations 41
   2.5 Sequences 48
   2.6 Maximum and Minimum 49
   2.7 Project: Another Approach 51

3 Integers   **52**
   3.1 Axioms 52
   3.2 Powers 55
   3.3 Characteristic Functions 57

4 Rational Numbers   **58**
   4.1 Axioms and Arithmetical Properties 58
   4.2 Powers 66
   4.3 Key Properties of $\mathbb{Q}$ 67

	4.4 Binomial Coefficients	67

5 Real Numbers   70
    5.1 Overview 70
    5.2 The Real Number System 71
    5.3 Arithmetical Operations 78
    5.4 The Archimedean Property 85
    5.5 Key Properties of the Reals 87
    5.6 The Extended Real Numbers 88

6 Complex Numbers   89
    6.1 The Complex Number System 89
    6.2 A Place to Stand 95

A Propositional Calculus   98
    A.1 Logical Operations 98
    A.2 Logical Rules 99

B Axioms   101
    B.1 Sets 101
    B.2 Natural Numbers 102
    B.3 Integers 103
    B.4 Rationals 103
    B.5 Reals 103
    B.6 Complex Numbers 104

C Consistency   105
    C.1 Mathematical Theories 105
    C.2 Sets and Relations 106
    C.3 Integers 108
    C.4 Rationals 108
    C.5 Reals 108
    C.6 Complex Numbers 108

D Redundancy   109
    D.1 Sets 109
    D.2 Natural Numbers 110
    D.3 Integers 110
    D.4 Rationals 111

	D.5 Reals	114
	D.6 Complex Numbers	114

E Solutions   **118**

F For the Teacher   **125**

Bibliography   **128**

Index   **129**

Preface

This book leads up to the starting-point of a rigorous course in basic real analysis. It is designed to satisfy a student who wants to start 'further back' than the axioms of a complete ordered field. Typically, such a student will be in the second or higher year at University, and will have attained some level of mathematical maturity.

We present a foundation in set theory, and build up through the natural numbers, integers, rational numbers, real and complex numbers, and we establish their properties on the basis of some more basic axioms.

We try to avoid various extremes:

- The theory is not a 'formal theory', in the strict, logical sense. There are no axioms for the logic used. It is thus a rigorous theory for actual people, not for automata.

- It does not try to do the impossible. You have to understand that, because of the work of twentieth-century logicians, we know that we cannot place mathematics on the kind of foundation of which David Hilbert dreamed. That is, it is impossible to give an account of analysis (or of any other sufficiently-rich mathematical theory) which is provably consistent. The only comfort I can offer is that if the theory is inconsistent, then it is already inconsistent by the end of the chapter about the natural numbers, Chapter 2.

- We do not get involved in the technical discussion of mathematical logic.

- We avoid doing 'clever' stuff that violates common-sense notions. For instance, we do not insist that every object under discussion be a set (or even a class). For instance, 2 is just 2, as opposed to something like $\{\emptyset, \{\emptyset\}\}$. There was a fashion for the latter at one stage.

Some of the ideas and concepts introduced are purely auxiliary, and you can forget about them once you have worked through this book. Indeed, the entire book is designed to be read in a week or two, and then forgotten, unless you are one of the people who want to dig deeper still, and get really serious about logic.

Where results are stated without proof, and if you are reading this book with a teacher, then your teacher may sometimes provide a proof. Otherwise you, the student, are expected to do so, as an exercise. The student should also note that some of the proofs given require elaboration. In reading the material, you should regard each sentence as a potential challenge: prove me if you can! Of course, unresolved difficulties should be brought to your instructor's attention. If, after fifteen minutes thought, you are making no progress, seek help. Some exercises are marked with a dagger (†). This means that there is a solution or a hint in Appendix E.

I would like to thank those Maynooth students and Erasmus programme visitors who helped find errors in drafts of this text, particularly Gareth Tracey and Gavin Armstrong. I am particularly grateful to my colleagues Detta Dickinson, Ann O'Shea and Stefan Bechtluft-Sachs, who read preliminary drafts and helped me avoid many errors. Any errors that remain are entirely my responsibility. If you find an error, please send the details to me at anthony.ofarrell@nuim.ie.

1: Sets and Relations

Except ye believe, ye shall not understand.[1]
— Isaiah vii 9

1.1 Overview

The theory of sets was invented by Georg Cantor (1845-1918). It formalises our intuitive notion of collections of things. Since the late nineteenth century, practically all mathematics has been formulated in the language of set theory.

In this chapter, we present an axiomatic account of set theory. More precisely, we begin to present such an account. Further axioms are presented at later points, as appropriate, and some of these have implications for set theory.[2].

Not everything will be a set. We start with arbitrary things. We do not define the term *thing*[3]. Some of our things will be rather abstract.

---

[1] Often attributed to St. Augustine of Hippo, who quotes and rephrases it in a sermon on the Gospel of John (Tractate XXIX, on John vii, 14-18). Without intending any blasphemy, one can say that this principle, which Augustine applies to religious belief, also applies to analysis. The student must be willing to believe that the whole thing makes sense, or she will never grasp it.

[2] For the *cognoscenti*, our theory of analysis is based on a set theory with atoms, and we take no position on the axiom of foundation or the continuum hypothesis. In short, it's a liberal set theory, that allows practically anything, the whole nine yards. Also, it is a formal theory with informal logic. We do not assume or use the axiom of choice in this book, nor is it needed in the elementary calculus course, but we have no great problem with its use in more advanced analysis courses

[3] You can't define everything.

1.2 Logic

We are going to use logic, but not worry about it. In other words, this is not a book *about* Logic. Nevertheless, a few words are in order.

1.2.1 Propositions

A character-string may be meaningless, or at least appear so. For instance:

Pitiless warfarin-stained piddle-less water-seekers
Prey not, pray now, soon pry less,
Prey not at all.

Some character-strings are proper sentences and mean something, but cannot be true or false. For instance, S.J. Perelman's essay title:

Sing softly, Fido, for I am sick with love.

Let us agree that we will deal with character strings that are meaningful, and are true or false. These are usually called propositions. We take it that there are such things, and that a proposition cannot be both true *and* false in the same context.

We make no attempt to give any procedure that you might use to recognise that something is a proposition, or to give any general procedure for deciding whether one is true. Any attempt to do that would turn this into a book about formal logic, and would also fail.

We do not distinguish between propositions, assertions and statements. As a result, we are not in a position to cope with issues such as the Liar Paradox[4], except by noting that it is not a proposition, nor is this sentence.

The elementary propositional calculus deals with the construction of compound propositions using the conjunction ('and': $p \wedge q$), disjunction ('inclusive or': $p \vee q$) and negation (not: $\neg p$) of propositions, and the truth tables that summarise the rules

---

[4] 'This is a lie.' If you would like to explore this further, I recommend J. Barwise and J. Etchemendy, *The Liar*, Oxford University Press, 1987.

1.2. LOGIC

governing the truth values (true or false) of compound propositions. This calculus is reviewed in Appendix A.

We recall that the equivalence $p \Leftrightarrow q$ means that p and q have the same truth value, i.e. $(p \wedge q) \vee (\neg p \wedge \neg q)$, and we define $p \Rightarrow q$ to mean $q \vee \neg p$.

In arguments that follow, we will make routine (and normally unacknowledged) use of the various rules of logic, such as the distributive law for *and* over *or*:

$$p \wedge (q \vee r) \Leftrightarrow (p \wedge q) \vee (p \wedge r).$$

These rules are established in Appendix A.

We shall make limited use of the formal symbols $\wedge$, $\vee$ and $\neg$, where clarity demands it, but for most of the book we shall just use the corresponding words (and, or, and not). This makes for a more readable text. However, the reader should note particularly that we always use 'or' in the inclusive sense, in contrast to ordinary English usage.

1.2.2 Equality

We take as meaningful the sentence '$a = b$.', whenever a and b are things. We don't define it, but we act as though it means that a and b are (names for) the same thing. We assume that the following are true propositions, for all things a, b and c:

$$\begin{aligned} a &= a. \\ (a = b) &\Rightarrow (b = a). \\ ((a = b) \wedge (b = c)) &\Rightarrow (a = c). \end{aligned}$$

We also assume that if $a = b$, then any proposition involving a is equivalent to the one obtained by replacing each occurrence of a by b [5].

We adopt the convention that := is to be read as 'is defined to be'.

---

[5] Thus, for instance, we have no option but to regard 'a is a different name for b' as a non-proposition! As you see, there are grounds for thinking that we *should* worry about logic; however we shall not.

1.2.3 Predicates, $\exists$ and $\forall$

We use the symbol $\exists$ to stand for 'there exists', and $\forall$ to stand for 'for all'. Thus if $\phi(x)$ is a predicate (i.e. $\phi(x)$ becomes a proposition when (the variable) x is replaced by (the name of) any thing), then
$$\exists x : \phi(x)$$
is true if and only if there is at least one thing a such that $\phi(a)$ is true, whereas
$$\forall x \; \phi(x)$$
is true if and only if $\phi(a)$ is true for each thing a. We also write this as
$$\phi(x), \forall x.$$

We recall that the negation of
$$\forall x \; \phi(x)$$
is
$$\exists x : \neg \phi(x)$$
whereas the negation of
$$\exists x : \phi(x)$$
is
$$\forall x \; \neg \phi(x).$$

For example, the negation of (All men are happy.) is (There exists an unhappy man.).

We use $\exists !$ to mean 'there exists a unique'. Thus
$$\exists ! x : \phi(x)$$
is true if and only if there is exactly one thing a such that $\phi(a)$ is true. For instance, anytime in AD 2009, it was true that
$$\exists ! \, x : x \text{ is a woman and is President of Ireland.}$$

For predicates $\phi(x)$ and $\psi(x)$, we say that $\phi(x) \Leftrightarrow \psi(x)$ if the propositions obtained by substituting anything for x are equivalent, and we define $\phi(x) \Rightarrow \psi(x)$ in a similar way.

We do not suggest how you might recognise a predicate.

1.3 Sets

1.3.1 Belonging: $\in$

We take as meaningful the sentence '*a belongs to A*', whenever a and A are any things. We use the alternative expressions '$a \in A$' or '*a is an element of A*' or '*is a member of A*' or '*A has a*' for 'a belongs to A'.

Some things are so horrible that they belong to nothing[6]. More precisely, there are things a such that $a \in X$ is false for all X. This is not necessarily obvious[7]. Indeed it can't be proven until we make a few more assumptions. Examples of such horrible things include those traditionally called *proper classes*. See Subsection 1.3.12 below.

1.3.2 Atoms

Some things may have no elements.

Definition 1.1. *An* atom *is a thing having no elements.*

(These are the atoms of set theory, not the atoms of chemistry). We are going to have atoms in our theory[8].

1.3.3 Classes

We take it that there is a kind of thing that is completely determined by its members (in other words, such things have no properties that are not determined by which elements they

---

[6] Lewis Carroll would have fun with that sentence. He once remarked that nothing is closer to ten than ten.

[7] However, consider Groucho Marx: "I wouldn't want to belong to any club that would have me as a member."

[8] In the theory of function spaces, a branch of Analysis used in Harmonic Analysis, Partial Differential Equations and Complex Analysis, there is another concept of atom, a special kind of function. The present atoms are sometimes referred to as 'urelements', in line with a practice of using the prefix ur- to indicate primitive objects of some type. We prefer the word atom, and regard it as unlikely that atomic functions and our atoms will occur in the same context, to any great extent.

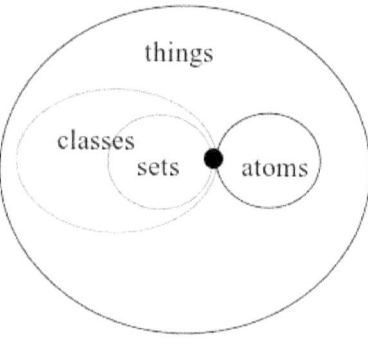

Figure 1.1: Kinds of Things

have). These things we call *classes*. We do not define them. We delineate the concept by means of axioms.

Not everything that has members is a class. The North Kildare Maths Problem Club is not a class, because it has properties that the class of its members could not have. For instance, it has a written constitution, has a bank account, and it sometimes lays on tea, coffeee, and biscuits. The first two axioms of set theory encapsulate the basic properties of classes.

1.3. SETS

The first excludes exotic members.

> **Axiom 1.1 (Membership).** *If A is a class, then each member of A is a class or an atom.*

The second makes precise the idea that all we know about a class is what its members are.

> **Axiom 1.2 (Extent, or Class Equality).** *If A and B are classes, then*
> $$(A = B) \Leftrightarrow (a \in A \Leftrightarrow a \in B).$$

Definition 1.2. *For classes A and B we say that $A \subset B$ (read as A is a subclass of B, or A is contained in B) if and only if*
$$a \in A \Rightarrow a \in B.$$

We observe that for classes, $A = B$ is equivalent to $A \subset B$ and $B \subset A$.

1.3.4 Sets

We distinguish between classes and sets. A set is a class that is not horrible:

Definition 1.3. *A set is a class A that belongs to some class B.*

Other collective nouns, such as *collection, family*, will be used as synonyms for class. We regularly use the word family for a class whose elements are all sets, as opposed to atoms.

So far, there could be no classes[9]. The third axiom provides a mechanism for constructing classes.

Axiom 1.3 (Classification). *For each predicate $\phi(x)$ there exists a class C such that*

$$(a \in C) \Leftrightarrow ((a \text{ is a set or an atom}) \text{ and } \phi(a)).$$

The Classification Axiom is sometimes called the Axiom of Comprehension. We denote this class C by $\{x : \phi(x)\}$. The sceptic will object that we have not told you how to recognise a predicate, so this axiom is not much use without that information. This is quite true, but we are avoiding logical formality.

The strings 'x is a set.' and 'x is an atom.' are predicates[10], so Axiom 1.3 allows us to use them, and related predicates, to define some specific classes, as follows:

Definitions 1.4. *We define the empty class:*

$$\emptyset := \{x : x \text{ is not a set or an atom}\},$$

the universal class:

$$\mathcal{U} := \{x : x \text{ is a set or an atom}\},$$

the class of all sets:

$$\mathcal{S} := \{x : x \text{ is a set}\},$$

the class of all atoms:

$$\mathcal{A} := \{x : x \text{ is an atom}\}.$$

---

[9] Karl Marx would be happy, but analysis would be very dull.
[10] This means that each thing is or is not a set, and each thing is or is not an atom.

1.3. SETS

So now we have at least one class, $\emptyset$.

Proposition 1. *$\emptyset$ is an atom and a set. No other atom is a class.*

Proof. Since $\emptyset$ is a class, the Axiom of Classification tells us that only sets or atoms may be members of it, and from its definition no set or atom is a member. Thus $\emptyset$ has no members, and hence is an atom.

Since $\emptyset$ is an atom, it belongs to the class $\mathcal{U}$. Thus it is a set, by the definition of *set*.

If an atom a were a class, then by the Axiom of Extent, it would equal $\emptyset$. $\square$

From the definition of $\mathcal{U}$, we see that $x \in \mathcal{U}$ is a shorthand way to say that x is an atom or a set.

1.3.5 Singletons

Definition 1.5. *Suppose a is a thing. Then 'singleton a' is the class*

$$\{a\} := \{x : x = a\}.$$

For instance, $\{\emptyset\}$ is the class whose only element is $\emptyset$, so it is different from $\emptyset$, by the Axiom of Extent. If a is a set or an atom, then a belongs to $\{a\}$, and nothing else belongs to it. If a is not a set or an atom, then $\{a\} = \emptyset$.

1.3.6 Operations on Classes

Definitions 1.6. *We define the union, intersection and difference of classes:*

$$\begin{aligned} A \cup B &:= \{x : (x \in A) \vee (x \in B)\}, \\ A \cap B &:= \{x : (x \in A) \wedge (x \in B)\}, \\ A \sim B &:= \{x : (x \in A) \wedge (x \notin B)\}. \end{aligned}$$

We say that A and B are disjoint *if $A \cap B = \emptyset$.*

These operations have the following properties:

Proposition 2. *Let A, B, and C be classes. Then:*

$$\begin{aligned}
A \cup (B \cup C) &= (A \cup B) \cup C, \\
A \cup B &= B \cup A, \\
A \cup \emptyset &= A, \\
A \cap (B \cap C) &= (A \cap B) \cap C, \\
A \cap B &= B \cap A, \\
A \cap \emptyset &= \emptyset, \\
A \cap (B \cup C) &= (A \cap B) \cup (A \cap C), \\
A \cup (B \cap C) &= (A \cup B) \cap (A \cup C), \\
A \sim (A \sim B) &= A \cap B, \\
A \sim (B \cup C) &= (A \sim B) \cap (A \sim C), \\
A \sim (B \cap C) &= (A \sim B) \cup (A \sim C).
\end{aligned}$$

Proof. These properties may be *intuitively understood* by drawing Venn diagrams. However, Venn diagrams are not proofs.

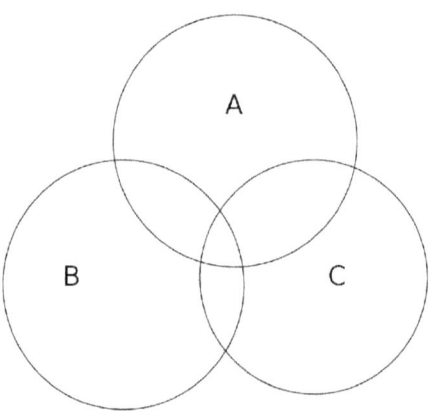

Figure 1.2: Venn Diagram

The properties may be *proved* by reducing them to logical propositions. The idea is to make the transition from the class A to the predicate $x \in A$. Since classes are determined by their elements, equality of classes reduces to equivalence of propositions about membership.

1.3. SETS

Take, for instance

$$A \cap (B \cup C) = (A \cap B) \cup (A \cap C).$$

This formula follows from the logical rule that

$$p \wedge (q \vee r) \Leftrightarrow (p \wedge q) \vee (p \wedge r).$$

The argument is:

$$\begin{aligned} x \quad &\in \quad A \cap (B \cup C) \\ &\Leftrightarrow \quad (x \in A) \wedge (x \in B \cup C) \\ &\Leftrightarrow \quad (x \in A) \wedge ((x \in B) \vee (x \in C)) \\ &\Leftrightarrow \quad ((x \in A) \wedge (x \in B)) \vee ((x \in A) \wedge (x \in C)) \\ &\Leftrightarrow \quad (x \in A \cap B) \vee (x \in A \cap C) \\ &\Leftrightarrow \quad x \in (A \cap B) \cup (A \cap C). \end{aligned}$$

Notice that this is not just a formal reduction. The logical rule may be proved by tabulating the 8 possible combinations of truth values of the propositions p, q and r. Thus proving a rule about (possibly infinite) classes reduces to checking 8 cases. □

Exercise 1.3.6.1. † Write out the proofs of all the remaining identities in Proposition 2.

1.3.7 Precedence

Expressions involving logical connectives and algebraic operators can become cluttered with brackets. To reduce this clutter, we use the rule that logical operators have lower precedence, i.e. are executed only after all algebraic operators have been evaluated.

1.3.8 Set Constructions

We make some assumptions that produce new sets from old.

Axiom 1.4 (Union). *If A and B are sets, then so is $A \cup B$.*

> **Axiom 1.5 (Power Set).** *If A is a set, then there is a set S such that*
> $$x \subset A \Leftrightarrow x \in S.$$

The set S is unique, by the Axiom of Extent.

1.3.9 Notation

We denote S by 2^A. It is called the *power set of A*.[11]

Definitions 1.7 (Arbitrary Unions and Intersections). *If $\mathfrak{A}$ is a family of sets[12], then*

$$\bigcup \mathfrak{A} := \bigcup_{A \in \mathfrak{A}} A := \{a : \exists A \in \mathfrak{A} \text{ such that } a \in A\}.$$

$$\bigcap \mathfrak{A} := \bigcap_{A \in \mathfrak{A}} A := \{a : a \in A, \forall A \in \mathfrak{A}\}.$$

> **Axiom 1.6 (Amalgamation).** *If $\mathfrak{A}$ is a set, then so is $\bigcup \mathfrak{A}$.*

Proposition 3.
(1) If A is a set, then each class contined in it is a set.
(2) If A is a set, then so is $\{A\}$.
(3) The intersection of an arbitrary nonempty family of sets is a set.
(4) $\bigcap \emptyset$ is equal to the universal class $\mathcal{U}$.

---

[11]Some books denote 2^A by $\mathcal{P}(A)$. The notation 2^A is more commonly used. See Section 3.3 for an account of the idea behind it.

[12]Recall that we normally use the word family for a class whose elements are all sets, as opposed to atoms.

1.3. SETS

Proof. (1) Follows from the Power Set Axiom.
(2) $\{A\}$ is a set for each set A, since it is a subset of 2^A.
(3) Follows from the axioms of Amalgamation and Power Set.
(4) is obvious from the definitions. □

If $A \subset B$ and B is a set, then we call A a *subset* of B.

We remark that by part (2), $2^\emptyset = \{\emptyset\}$ is a set, our second specific set.

Proposition 4 (De Morgan rules). *Let $\mathfrak{A}$ be a family of sets, and let B be a set. Then*
(1) $B \sim \bigcup \mathfrak{A} = \bigcap \{B \sim A : A \in \mathfrak{A}\}$;
(2) $B \sim \bigcap \mathfrak{A} = \bigcup \{B \sim A : A \in \mathfrak{A}\}$. □

Definition 1.8. *If a and b are each a set or atom, then the doubleton $\{a, b\}$ is defined by:*

$$\{a, b\} := \{a\} \cup \{b\}.$$

One readily checks that

$$2^{\{\emptyset\}} = \{\emptyset, \{\emptyset\}\}.$$

Thus by the Power Set Axiom, we have another set, which is clearly distinct from $\emptyset$ and from $\{\emptyset\}$.

Continuing in this way, we get many distinct sets. In particular, we shall henceforth take as well-defined such sets as $\{a, b, c\}$, $\{a, b, c, d\}$, and so on, where the elements are any sets or atoms.

1.3.10 Disjoint Unions

1.3.11 Notation

We use the symbol $\dot{\cup}$ to stand for 'disjoint union'. As well as denoting the union, it also functions as a kind of predicate. If we refer to $A \dot{\cup} B$, we are implicitly *asserting* that $A \cap B = \emptyset$. Similarly, if we refer to

$$\dot{\bigcup} \mathcal{A},$$

we are implicitly asserting that the sets belonging to the family $\mathcal{A}$ are pairwise-disjoint, i.e. that

$$A \cap B = \emptyset, \ \forall A, B \in \mathcal{A}.$$

1.3.12 Proper Classes

Definition 1.9. *A* proper class *is a class that is not a set.*

Proposition 5. *If A is a set, then A has a subset C that is not an element of A.*

Proof. Consider the class

$$C = \{x : x \in A \text{ and } x \notin x\}.$$

C exists by the Classification Axiom, and $C \subset A$, so C is a set, by the Power Set Axiom. It follows that

$$C \in C \Rightarrow C \notin C,$$

so $C \notin C$. Thus $C \notin A$. □

If you have read about Russell's Paradox, you will note that this argument is somewhat reminiscent of it, and that this proposition finesses the paradox. The following result avoids other 'antinomies' of naive set theory.

Proposition 6. *$\mathcal{U}$ and $\mathcal{S}$ are proper classes.*

Proof. If $\mathcal{S}$ were a set, then by Proposition 5 there would be a subset $C \subset \mathcal{S}$ that was not an element of $\mathcal{S}$, and hence not a set, contradicting the Power Set Axiom. Thus $\mathcal{S}$ is not a set.

It follows from the Power Set Axiom that the larger class $\mathcal{U}$ cannot be a set, either. □

Proposition 7. *If a is an atom and belongs to some set A, then $\{a\}$ is a set.* □

1.4. RELATIONS

1.3.13 Notation: $\exists x \in A$ **and** $\forall x \in A$

Definition 1.10. *If we have an expression $\phi(x)$ that becomes a proposition whenever x is replaced by any element of a class A (a 'predicate on A'), then we define the predicate $\phi'(x)$ to be*

$$\begin{cases} \phi(x), & \text{if } x \in A, \\ x \neq x, & \text{if } x \notin A. \end{cases}$$

We then define
$$\exists x \in A : \phi(x)$$
to mean
$$\exists x : \phi'(x).$$

We declare that
$$\forall x \in A, \phi(x)$$
or, equivalently,
$$\phi(x), \forall x \in A$$
is true if and only if
$$x \in A \Rightarrow \phi(x).$$

Similarly, we define
$$\{x \in A : \phi(x)\} = \{x : (x \in A) \wedge \phi'(x)\}.$$

If the expression $\phi(X)$ becomes a proposition whenever X is replaced by any subset of a set A, then we define
$$\{X \subset A : \phi(X)\} = \{X : (X \in 2^A) \text{ and } \phi'(X)\}.$$

1.4 Relations

1.4.1 Ordered Pairs

Here come our first non-class atoms.

> **Axiom 1.7 (Ordered Pairs).** *For each $a \in \mathcal{U}$ and $b \in \mathcal{U}$, there exists an atom (a, b), such that*
>
> $$(a, b) = (c, d) \Leftrightarrow ((a = c) \wedge (b = d)).$$
>
> *whenever c and d are sets or atoms.*

We call (a, b) an *ordered pair*[13].

Note that the ordered pair (a, b) coincides with (b, a) if and only if $a = b$.

1.4.2 Cartesian Product

Definition 1.11. *If A and B are classes, then the* Cartesian Product $A \times B$ *is the class of ordered pairs*

$$\{(a, b) : a \in A \ \wedge \ b \in B\}.$$

For example, we may consider $E \times F$, where E is the class of English words and F is the class of French words. Then $E \times F$ has, for example, (my, chien), (yes, peut) and (mauve, mauve).

We observe that the class of *all* ordered pairs is just $\mathcal{U} \times \mathcal{U}$.

1.4.3 Relations

Definitions 1.12. *A* relation *is a subclass of $\mathcal{U} \times \mathcal{U}$, i.e. a class of ordered pairs.*

If $r \subset A \times B$, then we call r a relation between elements of A and B. If $A = B$, we call r a relation between elements of A, and a relation on A.

---

[13] (a, b) is a new thing associated to a and b, not a class, but an 'atom-with-structure'.

1.4. RELATIONS

1.4.4 Examples

1. Let B denote the class of boys and G the class of girls[14]. The class
$$L = \{(x, y) \in B \times G : x \text{ loves } y\}$$
is a relation, which we may call the relationship relation.

2. Let P denote the class of all people. The class
$$B = \{(x, y) \in P \times P : x \text{ and } y \text{ are first cousins}\}$$
is a relation, the consanguinity relation.

3. Let S denote the class of items currently stocked by the SU shop and P be the class of possible prices, in euro. Then
$$p = \{(x, y) \in S \times P : y \text{ euro is the price of } x\}$$
is a relation, the SU-price relation.

4. Let E be the set of English words, and F be the set of French words. Then the class of *homonyms*
$$\{(x, y) \in E \times F : x = y\},$$
the class of *synonyms*
$$\{(x, y) \in E \times F : x \text{ means the same thing as } y\},$$
and the class of *cognates*
$$\{(x, y) \in E \times F : x \text{ derives from the same root as } y\},$$
are relations between elements of E and F.

---

[14] For the purposes of this and other such exercises, we assume that each person, and indeed each real object in the world, is an atom. This addition to the class of atoms is useful for exercises, but we do not actually need such atoms for the central results of real analysis. We like to think that real analysis would be the same, whether or not there are any people in the world, or any world.

5. Let
$$E = \{(a,b) : a \in b\}.$$
Then E is the membership relation on $\mathcal{U}$.

6. Let r be any relation. Then so is
$$r^{-1} = \{(y,x) : (x,y) \in r\}.$$
This is called the *inverse* of the relation r.
Note that $(r^{-1})^{-1} = r$.

There are three kinds of relation that are most important: *functions, equivalence relations, and partial orders*. We consider each in turn.

1.5 Functions

1.5.1 The Concept

Definitions 1.13. *A relation f is a* function[15] *if for each $x, y, z \in \mathcal{U}$ we have*
$$\left\{ \begin{array}{ll} (x,y) & \in f \\ (x,z) & \in f \end{array} \right\} \Rightarrow y = z.$$

If f is a function, and $(x,y) \in f$, we denote y by $f(x)$.
The domain *of f is*
$$\{x : (\exists y \in \mathcal{U} : (x,y) \in f)\}.$$
We denote it $\mathrm{dom} f$. *If $A = \mathrm{dom} f$, we say that f is a function on A. If $x \notin \mathrm{dom} f$, we say that $f(x)$ is undefined.*
The image *of f (or range of f) is*
$$\{y \in \mathcal{U} : (\exists x \in A : (x,y) \in f)\} = \{f(x) : x \in \mathrm{dom} f\},$$
and we denote it by $\mathrm{im} f$.

---

[15] In the Computer Science literature functions are sometimes referred to as *partial functions*. One finds sentences such as "This f is a partial function on the set A, but not a function". In our language, they would say: This function has as domain a proper subset of A, or this function is undefined except on a proper subset of A.

1.5. FUNCTIONS

Exercise 1.5.1.1. † For each of the examples of relations in Subsection 1.4.4 discuss whether or not it is a function. If the relation is a function, what are its domain and image?

1.5.2 Notation

When $f \subset A \times B$ is a function with domain A, we write $f : A \to B$ (read as "f maps A into B"). We will not use the notation $f : A \to B$ unless f is a function.

In various branches of Mathematics, one often finds other words used as synonyms for the word function. In particular, functions are sometimes called *maps*, as in the phrase: f is a map from A into B.

1.5.3 Notation

We denote the set of all functions from a set A into a set B by B^A.[16] For instance (anticipating some notation from Chapter 2), the set

$$\{0,1\}^{\{0,1,2\}}$$

has 8 elements. Thus if B has 2 elements and A has 3, then B^A has 2^3 elements. This, and its obvious generalisations, make the notation seem reasonable.

1.5.4 Axiom of Substitution

Intuitively, sets are 'small' classes. In keeping with this, we assume the following:

Axiom 1.8 (Substitution). *If f is a function on a set A, then $\operatorname{im} f$ is also a set.*

---

[16] See Section 3.3 for remarks about the relation of this notation to 2^A.

It follows that f itself is also a set, because we may define a new function g from A onto f by setting

$$g(x) = (x, f(x)), \forall x \in A.$$

Proposition 8. *Suppose that A and B are sets. Then so are $A \times B$ and A^B.*

Proof. Let $y \in B$. Then the formula

$$f(x) = (x, y), \ \forall x \in A,$$

defines a function on A, so

$$\text{im} f = A \times \{y\}$$

is a set, by the Axiom of Substitution.
Next,

$$g(y) = A \times \{y\}, \ \forall y \in B,$$

defines a function on B, so

$$\text{im} g = \{A \times \{y\} : y \in B\}$$

is also a set, by Substitution, and

$$\bigcup \text{im} g = A \times B$$

is then a set, by the Axiom of Amalgamation.
Finally,

$$A^B \subset 2^{A \times B}$$

is also a set, by the Power Set Axiom. $\square$

Exercise 1.5.4.1. Show that

$$\text{first} = \{((x, y), x) : x \in \mathcal{U}, y \in \mathcal{U}\}$$

is a function, with domain $\mathcal{U} \times \mathcal{U}$ and image $\mathcal{U}$.

Exercise 1.5.4.2. † Show that the class $\mathcal{U} \times \mathcal{U}$ of all ordered pairs is not a set.

1.5. FUNCTIONS

Exercise 1.5.4.3. † Show that the membership relation E of Example 5 of (1.4.4) is not a set.

Definitions 1.14. *1. The* identity function $1_A : A \to A$ *is the relation* $\{(x,x) : x \in A\}$. *This object is also called the* diagonal *of $A \times A$.*

2. If $f : A \to B$, and $E \subset A$, then the restriction of f to E *is the function*
$$f|E := \{(x, f(x)) : x \in E\}.$$

3. We denote the image of $f|E$ by $f(E)$[17]. *In other words,*
$$f(E) := \{f(x) : x \in E\}.$$

4. If f is a function on A, and $E \subset A$, and $g = f|E$, then we call f an extension *of g from E to A.*

Definitions 1.15. *A function $f : A \to B$ is* one-to-one, *or 1-1, or* injective, *if*
$$\forall (x,y) \in A \times A, \ f(x) = f(y) \Rightarrow x = y.$$

In this case, we also say that f is an injection. *A function $f : A \to B$ is* onto *B if $B = \text{im} f$. A function from A to B that is $1-1$ and onto B is said to be a* bijection *from A to B.*

If $f : A \to B$ and $g : B \to C$, then the composition $g \circ f$ *is*
$$\{(x,z) : \exists y \text{ such that } (x,y) \in f \land (y,z) \in g\}$$
$$= \{(x,z) : z = g(f(x))\}$$

---

[17]This notation risks potential ambiguity: It could happen that E is an element of A as well as being a subset of A. People occasionally use the notation $f[E]$ instead of $f(E)$, when this ambiguity is likely to arise. However, in most ordinary analysis this ambiguity will not arise, and it will be clear from context whether $f(p)$ represents a value or a set of values. We normally reinforce the distinction by using capitals for the names of sets of numbers or vectors and lower–case letters for their elements, although this is not always convenient. It is also common to use capitals for 'constant' numbers and vectors. You have probably noticed that in practice a 'constant' is a variable that doesn't change quite as quickly as the other ones.

i.e.
$$(g \circ f)(x) = g(f(x)), \ \forall x \in A.$$
If $f : A \to B$ and $g : B \to A$ are such that
$$g(f(x)) = x \ \forall x \in A,$$
and
$$f(g(y)) = y \ \forall y \in B,$$
then we say that *g* is an inverse function *for f*.

In some branches of Mathematics, bijections are called *tranformations*.

Proposition 9. *Let $f : A \to B$.*
(1) If g is an inverse function for f then f is an inverse for g.
(2) f has at most one inverse function.
(3) f has an inverse function if and only if f is 1-1 and onto.

Proof. (1) is obvious.
(2) Suppose $g : B \to A$ and $h : B \to A$ are inverses for f. Then for each $y \in B$,
$$g(y) = h(f(g(y))) = h(y),$$
hence $g = h$.
(3) Suppose f has an inverse, g. Suppose $f(x) = f(z)$. Then
$$x = g(f(x)) = g(f(z)) = z.$$
Thus f is 1-1.

Let $y \in B$ be given. Then $f(g(y)) = y$. Thus f is onto.

Conversely, suppose f is 1-1 and onto. Fix $y \in B$. Since f is onto, there exists $x \in A$ with $f(x) = y$. Since f is 1-1, this x is unique. Define $g(y) = x$.

Claim: g is an inverse for f.

1.5. FUNCTIONS

For each $y \in B$, we have from the definition of g that $f(g(y)) = y$.
It remains to show that $g(f(x)) = x$ for each $x \in A$. Fix $x \in A$ and let $y = f(x)$. From the definition of g we have $x = g(y)$. Thus
$$x = g(y) = g(f(x)).$$
This proves the claim and concludes the proof. $\square$

1.5.5 Notation

We denote the inverse function of f by f^{-1}, wherever it exists.

1.5.6

In the following exercises, suppose $f : A \to B$ is a function. Let $A_1, A_2 \subset A$.

Exercise 1.5.6.1. Show that $f(A_1 \cup A_2) = f(A_1) \cup f(A_2)$.

Exercise 1.5.6.2. † Give an example in which $f(A_1 \cap A_2) \neq f(A_1) \cap f(A_2)$.

Exercise 1.5.6.3. Suppose f^{-1} exists. Show that $f(A_1 \cap A_2) = f(A_1) \cap f(A_2)$.

Exercise 1.5.6.4. Let $f : A \to B$, $g : B \to C$, and $h : C \to D$. Show that $f \circ (g \circ h) = (f \circ g) \circ h$.

Exercise 1.5.6.5. † Give an example of $f : A \to A$ and $g : A \to A$ such that $f \circ g \neq g \circ f$.

Exercise 1.5.6.6. Suppose $f : A \to B$ and $g : B \to C$ are functions. Show that if $g \circ f$ is injective, then so are f and g.

Exercise 1.5.6.7. Suppose $\tau : A \to A$ is a function with $\tau \circ \tau = \mathbf{1}_A$ (*an involution,* so-called). Show that τ is a bijection from A onto A.

1.6 Equivalence Relations

1.6.1 The Concept

Definition 1.16. *Let A be a class. A relation $r \subset A \times A$ is an equivalence relation on A if the following three properties hold for all $x, y, z \in A$:*

(1) $\qquad\qquad (x,x) \in r, \ \forall x \in A;$ *(r is reflexive)*

(2) $\qquad\qquad (x,y) \in r \Rightarrow (y,x) \in r;$ *(r is symmetric)*

(3) $\qquad \left. \begin{array}{l} (x,y) \in r \\ (y,z) \in r \end{array} \right\} \Rightarrow (x,z) \in r.$ *(r is transitive)*

If r is an equivalence relation on A, we usually denote '$(x,y) \in r$' by '$x \approx y$' or some such notation, and we define the equivalence class *of x by*

$$[x] = \{y \in A : x \approx y\}.$$

1.6.2 Examples

1. The equality relation

    $$\{(x,x) : x \in \mathcal{U}\}$$

    is an equivalence relation on $\mathcal{U}$. Each singleton is an equivalence class.

2. The trivial relation

    $$\{(x,y) : x \in \mathcal{U}\}$$

    is another. There is just one equivalence class, the whole of $\mathcal{U}$.

1.6. EQUIVALENCE RELATIONS

If $\phi(x, y)$ is a predicate, and
$$\{(x, y) \in A \times A : \phi(x, y)\}$$
is an equivalence relation, then we call it *the equivalence relation on A defined by*
$$x \approx y \Leftrightarrow \phi(x, y).$$

Proposition 10. *If r is an equivalence relation on A, then*

(1) $$[x] \cap [y] \neq \emptyset \Rightarrow [x] = [y]$$

and

(2) $$A = \bigcup_{x \in A} [x].$$

Proof. (1) Suppose $[x] \cap [y] \neq \emptyset$, and pick $z \in [x] \cap [y]$. Then $x \approx z$ and $y \approx z$. By symmetry, $z \approx x$. By transitivity, $y \approx x$.

Let $w \in [x]$. Then $x \approx w$. By transitivity, $y \approx w$, hence $w \in [y]$. Thus $[x] \subset [y]$. Similarly, $[y] \subset [x]$, hence $[x] = [y]$.

(2) $[x] \subset A$ for each $x \in A$, hence
$$\bigcup_{x \in A} [x] \subset A.$$

Conversely, let $y \in A$. Then $y \in [y]$ by reflexivity, hence $y \in \bigcup_{x \in A} [x]$. Thus
$$A \subset \bigcup_{x \in A} [x].$$
Thus
$$A = \bigcup_{x \in A} [x].$$
□

Exercise 1.6.2.1. Suppose that $f : A \to B$. Show that
$$x \approx y \Leftrightarrow f(x) = f(y)$$
defines an equivalence relation on A.

The equivalence classes are usually called the *level sets* of f.

1.7 Partial Orders

1.7.1 The Concept

Definition 1.17. *Let A be a class. A relation $r \subset A \times A$ is a partial order on A if and only if the following three properties hold for all $x, y, z \in A$:*

(1) $\qquad (x, x) \in r \ \forall x \in A; \qquad$ *(reflexive)*

(2) $\qquad \left. \begin{array}{l} (x,y) \in r \\ (y,x) \in r \end{array} \right\} \Rightarrow x = y; \qquad$ *(antisymmetric)*

(3) $\qquad \left. \begin{array}{l} (x,y) \in r \\ (y,z) \in r \end{array} \right\} \Rightarrow (x,z) \in r. \qquad$ *(transitive)*

If r is a partial order, we usually write $(x, y) \in r$ as $x \preceq y$, or some such notation.

If $\phi(x, y)$ is a predicate, and

$$\{(x, y) \in A \times A : \phi(x, y)\}$$

is a partial order, then we call it *the partial order $\preceq$ on A defined by*

$$x \preceq y \Leftrightarrow \phi(x, y), \ \forall x, y \in A.$$

1.7.2 Examples

1. The relation defined by

   $$x \preceq y \Leftrightarrow x \subset y$$

   is a partial order on the class $\mathcal{S}$ of all sets.

2. The equality relation is a partial order on $\mathcal{U}$.

3. The trivial relation $A \times A$ on A is not a partial order on A unless A is a singleton, or empty.

1.7.3 Total Order

Definition 1.18. *A partial order r on A is a total order if*
$$((x,y) \in r) \vee ((y,x) \in r), \ \forall (x,y) \in A \times A.$$

Whenever we are given a total order $\preceq$ on a class A, we may define associated max and min operations:

$$\max(x,y) := \begin{cases} x, & \text{if } y \preceq x, \\ y, & \text{if } x \preceq y, \end{cases}$$
$$\min(x,y) := \begin{cases} y, & \text{if } y \preceq x, \\ x, & \text{if } x \preceq y. \end{cases}$$

Then
$$\min(x,y) = \min(y,x) \preceq x \preceq \max(x,y) = \max(y,x).$$

1.7.4 Strict Orders

A strict order on a set A is a relation r on A that is *antireflexive*:
$$(x,x) \notin r, \ \forall x \in A,$$
and is such that $r \cup \mathbf{1}_A$ is a total order on A.

One obtains a strict order from a total order by removing the diagonal $\mathbf{1}_A$ from it.

If r is a strict order, then given distinct $x, y \in A$, one has either xry or yrx, never both. Also xry and yrz imply xrz.

1.7.5 Upper and Lower Bounds

Definitions 1.19. *Suppose that $\preceq$ is a partial order on a class A, $B \subset A$, and $u \in A$. Then we say that u is an* upper bound *for B (with respect to $\preceq$) if and only if*
$$x \preceq u, \ \forall x \in B.$$

Similarly, we say that an element $b \in A$ is a lower bound *for B (with respect to $\preceq$) if and only if*
$$b \preceq x, \ \forall x \in B.$$

An element $m \in A$ is a least element of B (with respect to $\preceq$) if $m \in B$ and m is a lower bound for B. An element $m \in A$ is a greatest element of B (with respect to $\preceq$) if $m \in B$ and m is an upper bound for B.

Proposition 11. *B can have at most one least element, and at most one greatest element.* □

Based on this, we extend the definition of max and min:

Definition 1.20. *If $m \in B$ is the greatest element of $B \subset A$ with respect to a total order $\preceq$, then we call m the maximum of B (with respect to $\preceq$), and we write $m = \max B$.
Similarly, we define $\min B$.*

Definition 1.21. *A total order $\preceq$ defined on a class A is said to well-order A if each nonempty subclass $B \subset A$ has a least element. We also say that such a $\preceq$ is a well-ordering of A.*

Exercise 1.7.5.1. Discuss each of the following relations in terms of the concepts defined above. For instance, is it an equivalence relation, and if so, what are the equivalence classes? Is it a partial order? Is it a total order? Is it a well-ordering? Is it a function, and if so, what are its domain and image, and is it injective? Does it have an inverse function, and if so, what is it?

1. $\{(x, y) : x \text{ is the father of } y\}$.
2. $\{(x, y) : y \text{ is the father of } x\}$.
3. $\{(x, y) : x \text{ is the only son of } y\}$.
4. $\{(x, y) : x \text{ is a brother of } y\}$.
5. $\{(x, y) : x \text{ and } y \text{ are objects of type } t\}$, for some built–in Maple data type t. (If you are not familiar with Maple, try this for your favourite programming language.)

Exercise 1.7.5.2. † What is $\{\mathcal{U}\}$?

Exercise 1.7.5.3. Give an example of three sets A_1, A_2, A_3, such that each pair has nonempty intersection, but $A_1 \cap A_2 \cap A_3 = \emptyset$.

2: The Natural Numbers

The theory of the natural numbers formalises our notion of counting. The first published axiomatic theories of the natural numbers appeared in the 1880's, and were due to Gottlob Frege (1848-1925) and Giuseppe Peano (1858-1932)[1].

2.1 The Peano Postulates

2.1.1 The Postulates

We assume the following[2]:

Axiom 2.1 (Natural Numbers). *There is a particular set* $\mathbb{N} \subset \mathcal{A}$ *of atoms*[3], *and a particular atom* $1 \in \mathbb{N}$.

We call the elements of $\mathbb{N}$ *natural numbers*.

---

[1] It's worth noting that Peano built on the foundation in logic laid by George Boole of University College, Cork.

[2] The consistency of this theory of the natural numbers cannot be proved. It can be proved if we assume the consistency of one or another version of axiomatic set theory. This involves introducing set-theoretic axioms such as *foundation* and *infinity*, that are no more transparent than the Peano axioms, to say the least. Given such axioms, one may model Peano arithmetic, based on the interpretation of 1 as $\{\emptyset\}$ and $s(n)$ as $n \cup \{n\}$. See [4].

This does not mean that you should think of 1 as $\{\emptyset\}$. Think of 1 any way you like. Note, however, that in our treatment 1 is *not* a set.

[3] You may assume that the natural numbers are not ordered pairs, if you wish. In fact, we encourage you to do so. However we take no formal position on this.

2.1. THE PEANO POSTULATES

One might suppose that this axiom is just saying that there exists some nonempty set, because given a nonempty set A, we could call it $\mathbb{N}$ and pick an element $a \in A$ and call it 1. This would be a misunderstanding. First of all, we do not need to assume there is a nonempty set, because that fact already follows from previous axioms (see p. 13). The main point here is that we say that $\mathbb{N}$ is a particular set and 1 is a particular element. This means that there is just one set called $\mathbb{N}$ in our entire theory, and just one atom called 1. These things are fixed, once and for all. To be sure, the axiom tells us little else about these objects (apart from the rather important fact that $\mathbb{N}$ is a set, not just a class), and we'll need more axioms about them. Here they are:

Axiom 2.2 (Successor). *There is a particular function $s : \mathbb{N} \to \mathbb{N}$ that is a bijection from $\mathbb{N}$ onto $\mathbb{N} \sim \{1\}$.*

Axiom 2.3 (Induction).
$$\left\{ \begin{array}{rcl} A & \subset & \mathbb{N} \\ 1 & \in & A \\ n \in A & \Rightarrow & s(n) \in A \end{array} \right\} \Rightarrow (A = \mathbb{N}).$$

We call the function s *successor*, and call $s(n)$ the successor of n.

2.1.2 Discussion

Peano's original version listed rather more axioms about numbers and the successor function. More modern language allows them to be compressed into three statements. You will appreciate that the Successor Axiom is quite complex. It implies:

- s is a function.
- $s(n) \in \mathbb{N}$, $\forall n \in \mathbb{N}$.
- $s(n) \neq 1$, $\forall n \in \mathbb{N}$.
- $\forall n \in \mathbb{N}$, if $n \neq 1$, then $\exists m \in \mathbb{N}$ with $s(m) = n$, and
- $s(n) = s(n') \Rightarrow n = n'$.

All this is wrapped up in the concept of a bijection. Notice also that s is a bijection from the set $\mathbb{N}$ onto a *proper subset of itself*. This is the characteristic property of what are called *infinite* sets:

Definition 2.1. *A set A is* infinite *if there is a bijection from A onto a proper subset of A. A set A is* finite *if it is not infinite.*

So the Successor Axiom, together with Axiom 2.1, tells us that there is at least one infinite set.

In your previous mathematical experience, you have probably already seen something of the power of the Axiom of Induction.

In what follows, We shall define the order $a \leq b$ and the arithmetic operations $a + b$, $a \times b$ and a^b on $\mathbb{N} \times \mathbb{N}$, and obtain the familiar properties of these relations[4]. We proceed to spell out the details.

2.1.3 Two, Three, etc

We start by naming a few natural numbers.

Definitions 2.2.

$$\begin{aligned} 2 &:= s(1), & 3 &:= s(2), & 4 &:= s(3), \\ 5 &:= s(4), & 6 &:= s(5), & 7 &:= s(6), \\ 8 &:= s(7), & 9 &:= s(8), & 10 &:= s(9). \end{aligned}$$

---

[4] Briefly, these properties may be summarised by saying that $\leq$ well-orders $\mathbb{N}$, $(\mathbb{N}, +)$ is a commutative cancellation semigroup without identity, $(\mathbb{N}, \times)$ is a commutative cancellation semigroup with identity 1, $\times$ distributes over $+$, and a^b satisfies the laws of exponents.

2.2 Ordering $\mathbb{N}$

From now on, we shall take it that the reader can supply (or at least imagine) the definition of any specific natural number, such as that represented in base 10 by 432231^5.

2.2 Ordering $\mathbb{N}$

In this section we define the standard order on the natural numbers. For this purpose we introduce the auxiliary concepts of final and initial sets[6]. Once they have served their purpose, you can forget about them.

2.2.1 Final and Initial Sets

Definitions 2.3. *A set $F \subset \mathbb{N}$ is* final *if it is nonempty and*

$$m \in F \Rightarrow s(m) \in F.$$

A set $I \subset \mathbb{N}$ is initial *if $\mathbb{N} \sim I$ is final.*

For instance, $\mathbb{N}$ is final and $\emptyset$ is initial.

Proposition 12.
(1) The union of two final sets is final. More generally, if $\mathcal{F}$ is a nonempty family of final sets, then $\bigcup \mathcal{F}$ is final.
(2) If $\mathcal{F}$ is a nonempty family of initial sets, then $\bigcap \mathcal{F}$ is initial.
(3) Let $I \subset \mathbb{N}$. Then the following are equivalent:

1. *I is initial.*

2. *$\forall m \ (m \in I) \vee (s(m) \notin I)$.*

3. *$\forall m \ m \notin I \Rightarrow s(m) \notin I$.*

---

[5]Incidentally, the fact that 1,2,3 and others have already been used to label chapters, propositions, equations and even pages does not affect the logical integrity of this account. One might think that they have been used before being defined. However, they have not been used *as natural numbers*, just as labels. We could have used any labels at all for these elements of the book, such as coloured flags or comic-book characters, or the names of plants, instead of numbers.
[6]For this idea we are indebted to Kelley [4]

4. $\forall m\ s(m) \in I \Rightarrow m \in I$.

(4) If F is final, then so is $s(F)$. *(5)* If F is final and I is initial, then $F \sim I$ is final. □

The following proposition gives alternative forms of the Axiom of Induction:

Proposition 13.
(1) If $1 \in F$ and F is final, then $F = \mathbb{N}$.
(2) If I is initial and nonempty, then $1 \in I$.
(3) If I is initial and $1 \notin I$, then $I = \emptyset$. □

Definitions 2.4. *Given $n \in \mathbb{N}$, let*
$$\begin{aligned} n^+ &:= \bigcap\{F : n \in F, \text{and } F \text{ is final}\}, \\ n^- &:= \mathbb{N} \sim n^+. \end{aligned}$$

Proposition 14. $1^+ = \mathbb{N}$, and hence $1^- = \emptyset$.

Proof. Use Proposition 13(1). □

Proposition 15. *Let $n \in \mathbb{N}$. Then*
(1) $n \in n^+$, so $n \notin n^-$,
(2) n^+ is final, so n^- is initial, and
(3) $\mathbb{N} = n^+ \cup n^-$. □

We are going to show that each final set is some n^+, but we have to do some preliminary work, first.

Proposition 16. *Let $n \in \mathbb{N}$. Then:*
(1) $s(n)^+ = s(n^+)$;
(2) $n^+ = \{n\} \cup s(n)^+$;
(3) $n \notin s(n)^+$.

Proof. (1): By Proposition 12(4), $s(n^+)$ is a final set. Also $s(n) \in s(n^+)$, so $s(n)^+ \subset s(n^+)$ from the definition of $\cdot^+$.

For the other direction, fix a final set F with $s(n) \in F$, and fix $m \in s(n^+)$. We need to show that $m \in F$.

Choose $r \in n^+$ with $m = s(r)$. Take $F_1 = F \cup \{n\}$. Then F_1 is final, so $n^+ \subset F_1$, so $r \in F_1$, so $r = n$ or $r \in F$, so

2.2. ORDERING ℕ

$m = s(n) \in F$ or $m \in F$, so $m \in F$, and we are done.

(2): The right-hand side is final, and n belongs to it, so it contains n^+, by definition.

For the other direction, let F be final and $n \in F$. We have to show that the right-hand side is contained in F, so we have to show that $s(n)^+ \subset F$. Let $m \in s(n)^+$. By part (1), there exists $r \in n^+$ with $m = s(r)$. Then $r \in F$, by definition of n^+, so $m \in F$, by definition of final.

(3): This one needs induction. Let

$$I = \{n \in \mathbb{N} : n \in s(n)^+\}.$$

Then $1 \notin I$, since that would imply $1 \in s(1)^+ = s(1^+)$ by Part (1), and yet $1 \notin s(\mathbb{N})$, a contradiction.

Next, we show that I is initial, by proving that

$$s(n) \in I \Rightarrow n \in I.$$

Suppose $s(n) \in I$. Then $s(n) \in s(s(n))^+ = s(s(n^+))$. Choose $m \in n^+$ such that $s(n) = s(s(m))$. Since s is one-to-one, we conclude that $n = s(m) \in s(n^+) = s(n)^+$, so $n \in I$.

By Proposition 13(3), $I = \emptyset$, and we are done. □

Corollary 17. *For each $n \in \mathbb{N}$, n^+ is the disjoint union of $\{n\}$ and $s(n)^+$, and hence $s(n)^-$ is the disjoint union of $\{n\}$ and n^-.* □

For instance, applying this a few times, we see that

$$5^- = \{1, 2, 3, 4\}.$$

Proposition 18.
(1) Let F be final, and $n \in \mathbb{N}$. Then $F \subset n^+$ or $n^+ \subset F$.
(2) Let I be initial, and $n \in \mathbb{N}$. Then $I \subset n^-$ or $n^- \subset I$.
(3) If $m, n \in \mathbb{N}$, then $m^+ \subset n^+$ or $n^+ \subset m^+$.

Proof. (1) Let $A = \{n \in \mathbb{N} : (n^+ \subset F) \vee (F \subset n^+)\}$. We shall use induction to show that $A = \mathbb{N}$.

First, $F \subset \mathbb{N} = 1^+$, so $1 \in A$.

Next, suppose $n \in A$. Then $n^+ \subset F$ or $F \subset n^+$.

If $n^+ \subset F$, then $s(n)^+ \subset F$.

If, on the other hand, $n^+ \not\subset F$ and $F \subset n^+$, then $n \notin F$, so $F \subset s(n)^+$, by Proposition 16(2).

Thus $s(n) \in A$, in any case.

By the Axiom of Induction, $A = \mathbb{N}$, and we are done. Parts (2) and (3) are immediate from Part (1). $\square$

Proposition 19.
(1) If $F \subset \mathbb{N}$ is final, then $F = n^+$ for some $n \in \mathbb{N}$.
(2) If $I \subset \mathbb{N}$ is initial, then $I = n^-$ for some $n \in \mathbb{N}$.

Proof. It suffices to prove (2).

Suppose, on the contrary, that I is initial and $I \neq n^-$ for all $n \in \mathbb{N}$.

Then $I \neq 1^- = \emptyset$, so $1 \in I$ by Proposition 13.

Suppose $n \in I$. Then $I \not\subset n^-$, so $n^- \subset I$, by Proposition 18. Thus $s(n)^- = \{n\} \cup n^- \subset I$.

Now suppose also that $s(n) \notin I$. Then $s(s(n))^- = \{s(n)\} \cup s(n)^- \not\subset I$, so $I \subset s(s(n))^-$. Thus I is trapped between two m^- sets that differ by a single element, and so equals one or the other, a contradiction.

Thus $n \in I \Rightarrow s(n) \in I$. By induction, $I = \mathbb{N}$, which is impossible for an initial set. $\square$

Proposition 20. *Let $m, n \in \mathbb{N}$. Then $m \neq n \Rightarrow n^+ \neq m^+$. Thus*
$$(m = n) \Leftrightarrow (n^+ = m^+) \Leftrightarrow (n^- = m^-).$$

Proof. We use a double induction argument:

Let
$$\begin{aligned} A_n &= \{m \in \mathbb{N} : m = n \vee m^+ \neq n^+\}, \\ A &= \{n \in \mathbb{N} : A_n = \mathbb{N}\}. \end{aligned}$$

First, if $m \neq 1$, then m belongs to the final set $s(\mathbb{N})$, which does not have 1, so $1 \notin m^+$, so $1^+ \neq m^+$. Thus $A_1 = \mathbb{N}$, so $1 \in A$.

Next, suppose $n \in A$, and consider $A_{s(n)}$.

We have $s(n) \neq 1$, so $1^+ \neq s(n)^+$, so $1 \in A_{s(n)}$.

2.2. ORDERING ℕ

Suppose $m \in A_{s(n)}$. Then $m \in A_n$, so since $m = n$ or $m^+ \neq n^+$, and s is injective, we conclude that $s(m) = s(n)$ or $s(m^+) \neq s(n^+)$, so by Proposition 16(1) $s(m) = s(n)$ or $s(m)^+ \neq s(n)^+$, so $s(m) \in A_{s(n)}$.

By the Axiom of Induction, $A_{s(n)} = \mathbb{N}$, so $s(n) \in A$.

So by another application of induction, $A = \mathbb{N}$, and we are done. □

2.2.2 Defining ≤, <, ≥, > on ℕ

See Section 1.7 for terminology about partial orders, strict orders and total orders.

Definition 2.5. *For $m, n \in \mathbb{N}$, let $m \leq n$ denote $m^- \subset n^-$, and let $m < n$ denote $m \leq n$ and $m \neq n$. We define $n \geq m$ to mean $m \leq n$ and $n > m$ to mean $m < n$.*

Proposition 21. *Let $m, n \in \mathbb{N}$. Then the following are mutually equivalent: (1) $m \leq n$. (2) $n^+ \subset m^+$. (3) $n \in m^+$.*

Proof. The equivalence of (1) and (2) is trivial. (2) implies (3) since $n \in n^+$.

To see (3) ⇒ (2), let $n \in m^+$. Then since m^+ is final, the definition of n^+ gives $n^+ \subset m^+$. □

Proposition 22. $n < s(n)$, $\forall n \in \mathbb{N}$.

Proof. This follows from Proposition 16(2). □

Proposition 23. *Let $m, n \in \mathbb{N}$. Then the following are mutually equivalent: (1) $m < n$. (2) $n^+ \subset s(m)^+$. (3) $n \in s(m)^+$.* □

Proposition 24. $n = \min n^+$, $\forall n \in \mathbb{N}$ □

Proposition 25. $\leq$ *is a total order on* $\mathbb{N}$.

Proof. $\subset$ is a partial order on sets. It follows that $\leq$ is reflexive and transitive. It is antisymmetric by Proposition 20. It is a total order by Proposition 18 □

Proposition 26. *Suppose $n \leq m \leq s(n)$. Then $m = n$ or $m = s(n)$.*

Proof. We have $s(n)^+ \subset m^+ \subset n^+$, so $m^+ = n^+$ or $m^+ = s(n)$, by Proposition 16(2). The result now follows from Proposition (20). □

Proposition 27. $\mathbb{N}$ *is well-ordered by $\leq$.*

Proof. Let $E \subset \mathbb{N}$ be nonempty.
  Let $F = \bigcup \{n^+ : n \in E\}$.
  Then F is a final set, so $F = m^+$ for some $m \in \mathbb{N}$, by Proposition 19.
  $n^+ \subset m^+$, for all $n \in E$, so $m \leq n$, $\forall n \in E$.
  $m \in F$, so $\exists n \in E$ with $m \in n^+$. But then $n \leq m$ by Proposition 21, so $m = n$. So $m \in E$.
  Thus E has a least element, namely m. □

Exercise 2.2.2.1. † Show that $m < n$ and $n \leq p$ imply $m < p$.

Exercise 2.2.2.2. Show $\geq$ is a total order on $\mathbb{N}$.

Exercise 2.2.2.3. Show $<$ and $>$ are strict orders on $\mathbb{N}$[7].

2.2.3 Notation

In later chapters, we shall often use the standard notation

$$\{n, n+1, \ldots\}$$

for n^+, and use $\{1, \ldots, n\}$ for $s(n)^-$. This is based on the fact that $s(n)^-$ consists of 1, n, and each $m \in \mathbb{N}$ with $1 < m < n$.

---

[7]cf. page 27.

2.3 Inductive Definitions

2.3.1 The General Version

Theorem 28. *Let A be a set,*

$$B = \bigcup_{n \in \mathbb{N}} \{n\} \times A^{(n^-)},$$

and let $g : B \to A$. Then $\exists! f : \mathbb{N} \to A$ such that

$$f(n) = g(n, f|(n^-)), \; \forall n \in \mathbb{N}.$$

2.3.2 Remark

The datum g is a function defined on the set

$$B = \{(n, h) : n \in \mathbb{N} \text{ and } h : (n^-) \to A\}.$$

An element of the domain of g is a pair (n, h), where the function h has the initial set n^- as domain and has values in A. Also, $g(n, h)$ must be defined for each such pair.

Note that when $n = 1$, $n^- = \emptyset$, so the only possible h is the empty function $\emptyset$, and $g(1, \emptyset)$ is the 'initial value' $f(1) \in A$. For other n, the formula in the statement 'defines $f(n)$ in terms of n and the previous values of f'. The expression $f|(n^-)$ denotes, of course, the restriction of the function f to the set n^-, and this contains precisely all the information about the previous values of f. From here on, we will usually abbreviate $f|(n^-)$ to $f|n^-$.

2.3.3

Proof of Theorem 28. Define f to be the union

$$\bigcup \left\{ h : \exists n \in \mathbb{N} : h \in A^{(n^-)}, \text{ and } h(m) = g(m, h|m^-), \; \forall m \leq n \right\}.$$

One checks, using induction, that if $h \in A^{(n^-)}$ and $k \in A^{(r^-)}$ satisfy
$$\begin{aligned} h(m) &= g(m, h|m^-), \; \forall m \leq n, \\ k(m) &= g(m, k|m^-), \; \forall m \leq r, \end{aligned}$$

then $h(m) = k(m)$, $\forall m < \min\{n, r\}$. Thus f is a function.

Again, using induction, one checks that each $n \in \mathbb{N}$ belongs to the domain of some h, so that the domain of f is $\mathbb{N}$. In detail, let C denote the set of those $n \in \mathbb{N}$ for which there exists some $h : n^- \to A$ with $h(m) = g(m, h|m^-)$, $\forall m \leq n$. We claim that $C = \mathbb{N}$.

First, $1 \in C$, since the empty function is such an h (and indeed the only one). Next, suppose $n \in C$, and choose $h_1 : n^- \to A$ such that $h_1(m) = g(m, h_1|m^-)$, $\forall m \leq n$. Define $h : s(n)^- \to A$ by setting $h(m) = h_1(m)$ whenever $m < n$ and $h(n) = g(n, h_1|n^-)$. Then clearly h has $h(m) = g(m, h|m^-)$, $\forall m \leq s(n)$. Thus $s(n) \in C$. Induction now gives the claim.

Finally, a third use of induction shows that f is uniquely-determined by the conditions of the statement. □

2.3.4 Versions

The statement of this theorem is rather abstract, its proof needs close attention, and it is probably the single most difficult passage in this book. However, this is one of the most useful theorems in all mathematics. The version just given is the 'complete induction' version, in which the value of n and of all previous values of the function may be used to define the value $f(n)$ at n. The single-step version is the commonest used:

Corollary 29. *Let A be a set, $a \in A$, and $g : \mathbb{N} \times A \to A$. Then $\exists! f : \mathbb{N} \to A$ such that $f(1) = a$ and*

$$f(s(n)) = g(n, f(n)), \ \forall n \in \mathbb{N}.$$

□

A special case is:

Corollary 30. *Let A be a set, $a \in A$, and $g : A \to A$. Then $\exists! f : \mathbb{N} \to A$ such that $f(1) = a$ and*

$$f(s(n)) = g(f(n)), \ \forall n \in \mathbb{N}.$$

□

In this version, only the *value* of $f(n)$ (and not the value of n) is used to define the next value $f(s(n))$. This process is usually referred to as 'iterating the function g', and the function f is said to be 'iteratively-defined'.

Finally, we note the complete-induction version in which the data do not explicitly depend on n, but only on the previous values of the function:

Corollary 31. *Let A be a set,*

$$B = \bigcup_{n \in \mathbb{N}} A^{(n^-)},$$

and let $g : B \to A$. Then $\exists! f : \mathbb{N} \to A$ such that

$$f(n) = g(f|(n^-)), \ \forall n \in \mathbb{N}.$$

Exercise 2.3.4.1. † Prove each of the three corollaries by defining, in each case, a suitable function g and appealing to the theorem.

Exercise 2.3.4.2. Write out direct proofs of the three corollaries, without appealing to the theorem.

2.4 Binary Operations

Definition 2.6. *Given a set A, a* binary operation on A *is a function $* : A \times A \to A$.*

The main examples are the arithmetic operations such as $(x, y) \mapsto x + y$, introduced below. When $*$ is a binary operation, one sometimes denotes $*(x, y)$ by the so-called *infix notation* $x * y$.

By analogy, the suggestive terminology *unary operation on A* is used for a function mapping $A \to A$. The main examples of this are 'minus' and 'reciprocal', introduced below.

We proceed to define the arithmetic operations on $\mathbb{N}$, one by one.

2.4.1 Addition

We are going to define $m + n$ for natural numbers m and n.

Definition 2.7. *Fix $m \in \mathbb{N}$. Then $m+n$ is defined for all n by the rules:*

$$\begin{aligned}&(1) \quad m + 1 := s(m), \\ &(2) \quad n = s(p) \Rightarrow m + n := s(m + p).\end{aligned}$$

Rule (2) may be rephrased: $m + s(p) := s(m + p)$, $\forall p \in \mathbb{N}$.

To reconcile this definition with the type sanctioned by Corollary 30, take $A = \mathbb{N}$, $a = s(m)$ and $g = s$. Then $m + n$ is $f(n)$, where f is the function whose existence and uniqueness is guaranteed by the corollary. Thus we may state:

Theorem 32. *$m+n$ is a well-defined natural number, for each $m, n \in \mathbb{N}$.* □

From now on, we shall normally refer to $s(m)$ by the more usual notation $m + 1$.

Next, we show that addition is associative and commutative.

Proposition 33 (+ is associative). *For each $m, n, r \in \mathbb{N}$ we have*

$$m + (n + r) = (m + n) + r.$$

Proof. We use induction on r.

First, consider $r = 1$. For all m and n, from the definition of $+$, we have

$$m + (n + 1) = m + s(n) = s(m + n) = (m + n) + 1.$$

Now suppose that for a given r we have $m + (n + r) = (m + n) + r$ for all $m, n \in \mathbb{N}$. Then for each m and n, we have

$$\begin{aligned}m + (n + s(r)) &= m + s(n + r) &&\text{[by definition of +]} \\ &= s(m + (n + r)) &&\text{[by definition of +]} \\ &= s((m + n) + r) &&\text{[by hypothesis]} \\ &= (m + n) + s(r). &&\text{[by definition of +]}\end{aligned}$$

Thus $m + (n + s(r)) = (m + n) + s(r)$ for all $m, n \in \mathbb{N}$. □

2.4. BINARY OPERATIONS

Henceforth, we write $m + n + p$, without brackets.

Lemma 34. $m + 1 = 1 + m$ for all $m \in \mathbb{N}$.

Proof. We use induction on m.

The case $m = 1$ is obvious.

Suppose it holds for m. Then using the definition of $+$ and this assumption, we get $s(m) + 1 = s(s(m)) = s(m+1) = s(1+m) = 1 + s(m)$. Thus it holds for $s(m)$. $\square$

Proposition 35 (+ is commutative).
$m + n = n + m$, $\forall m, n \in \mathbb{N}$.

Proof. We use induction on n.

It holds for all m with $n = 1$, by the lemma.

Suppose it holds for all m and a certain n. Then using this hypothesis, the definition of $+$, the lemma, and associativity, we get that for each m, $m + s(n) = s(m+n) = s(n+m) = n + s(m) = n + (m+1) = n + (1+m) = (n+1) + m = s(n) + m$. Thus it holds for $s(n)$ as well. $\square$

Proposition 36 (Cancellation Law). *If m, n, $p \in \mathbb{N}$ and $m + p = n + p$, then $m = n$.*

Proof. We use induction on p to prove the proposition

$$\Phi(p): \quad (m \subset \mathbb{N}) \wedge (n \in \mathbb{N}) \wedge (m + p = n + p) \Rightarrow m = n.$$

$\Phi(1)$ is true, since s is injective.

Suppose $\Phi(p)$, and let $m + (p+1) = n + (p+1)$. Then (using associativity and commutativity) $(m+1) + p = m + (1+p) = m + (p+1) = n + (p+1) = n + (1+p) = (n+1) + p$, so (by the hypothesis) $m + 1 = n + 1$, so (since s is injective) $m = n$. Thus $\Phi(p+1)$ holds. $\square$

Proposition 37. *If $m, n, p \in \mathbb{N}$, then $m \leq n \Rightarrow m + p \leq n + p$,*

Proof. We use induction on p.

Case $p = 1$: Let $m \leq n$. Then $n^+ \subset m^+$, so $s(n)^+ = s(n^+) \subset s(m^+) = s(n^+)$, so $s(m) \leq s(n)$, i.e. $m + 1 \leq n + 1$.

Step: Suppose it holds for p, and let $m \leq n$. Then $m + p \leq n + p$, so by Case 1, $m + (p+1) = (m+p) + 1 \leq (n+p) + 1 = n + (p+1)$. □

These two propositions immediately yield:

Corollary 38. *If $m, n, p \in \mathbb{N}$, then $m < n \Rightarrow m + p < n + p$* □

2.4.2 Subtraction

We are going to define $n - m$ for natural numbers $m < n$.

Proposition 39. *Let $n \in \mathbb{N}$. Then $\{n + m : m \in \mathbb{N}\} \subset s(n)^+$.*

Proof. One sees by induction on m, using the definition of $+$ and the fact that $s(n)^+$ is final, that $n + m \in s(n)^+$ for each $m \in \mathbb{N}$. □

(Soon we will see that the reverse inclusion holds.) Combining this with Proposition 16(3) we get:

Proposition 40. *If $n, m \in \mathbb{N}$, then $n \neq n + m$.* □

The following is a variant of the principle of induction, corresponding to induction 'starting at a certain number'.

Proposition 41. *Let $m \in \mathbb{N}$ and $A \subset \mathbb{N}$. Then*

$$\left. \begin{array}{rcl} m & \in & A \\ A & \subset & m^+ \\ n \in A & \Rightarrow & n + 1 \in A \end{array} \right\} \Rightarrow A = m^+.$$

Proof. Suppose $m \in A \subset m^+$ and $m \in A \Rightarrow m + 1 \in A$. Take $B = A \cup m^-$, and use induction to show $B = \mathbb{N}$. (For the initial step, treat the cases $m = 1$ and $m > 1$ separately. For the induction step from n to $n + 1$, consider separately the cases $n < m$ and $n \geq m$.) □

2.4. BINARY OPERATIONS

Proposition 42. *Let $m, n \in \mathbb{N}$, and $m < n$. Then $\exists! r \in \mathbb{N}$ such that $m + r = n$.*

Proof. Fix $m \in \mathbb{N}$, and take

$$A = \{n > m : \exists r \in \mathbb{N} \text{ with } m + r = n\}.$$

Then $s(m) \in A$, and $n \in A \Rightarrow s(n) \in A$, so Proposition 41 gives $A = s(m)^+$. This gives the existence part.

The uniqueness follows from the Cancellation Law, Proposition 36. $\square$

Definition 2.8. *When $m, n \in \mathbb{N}$ and $m < n$, then we define $n - m$ to be the unique $r \in \mathbb{N}$ such that $m + r = n$.*

Proposition 43. *If $m, n, p \in \mathbb{N}$ and $m < n < p$, then*

$$p - m = (p - n) + (n - m).$$

Proof. Let $r = n - m$ and $s = p - n$. Then $p = n + s = m + r + s$, so $p - m = r + s$, as required. $\square$

Exercise 2.4.2.1. † Show that $5 - 2 = 3$.

2.4.3 Multiplication

We define the product $m \times n$ of two natural numbers:

Definition 2.9. *Fix $m \in \mathbb{N}$. Then $m \times n$ is defined by the rules:*

$$\begin{aligned}(1) \quad & m \times 1 := m, \\ (2) \quad & m \times (p+1) := (m \times p) + m, \; \forall p \in \mathbb{N}.\end{aligned}$$

Proposition 44. *$m \times n$ is well-defined, for each $m, n \in \mathbb{N}$.*

Proof. This follows on applying Corollary 30 with $A = \mathbb{N}$, $a = m$, and $g : n \mapsto n + m$. $\square$

The reader is invited to prove the following, using induction:

Proposition 45 (Associative and Commutative Laws).
For each $m, n, r \in \mathbb{N}$ we have

$$\begin{aligned}
(1) \quad m \times (n \times r) &= (m \times n) \times r, \\
(2) \quad m \times n &= n \times m.
\end{aligned}$$

$\square$

Proposition 46 (Cancellation Law). *If $m, n, p \in \mathbb{N}$ and $m \times p = n \times p$, then $m = n$.* $\square$

Proposition 47. $m \leq n \Rightarrow m \times p \leq n \times p$, $\square$

Proposition 48. *Let $m, n \in \mathbb{N}$. then $m \leq m \times n$, with equality only if $m = 1$ or $n = 1$.* $\square$

Proposition 49 (Distributive Law). *Let $m, n, p \in \mathbb{N}$. Then*

$$m \times (n + p) = (m \times n) + (m \times p).$$

$\square$

Henceforth we usually write mn or $m \cdot n$ for $m \times n$, and we write products such as mnp without brackets.

We can now define the meaning of expressions such as x is a factor of y, y is divisible by x, x is even, and so on. We omit the details, but we shall use these terms whenever convenient.

Exercise 2.4.3.1. † Show that $n + 1 \leq 2n$, for each $n \in \mathbb{N}$.

2.4.4 Factorial

The factorial $n!$ of a natural number n is defined inductively by the rules $1! = 1$ and $(n+1)! = (n+1) \times (n!)$.

2.4. BINARY OPERATIONS

2.4.5 Powers

We define the power m^n for natural numbers m and n:

Definition 2.10. *Fix $m \in \mathbb{N}$. Then m^n is defined by the rules:*

(1) $m^1 := m$,
(2) $m^{(p+1)} := (m^p) \times m$.

Proposition 50. *For each $m, n \in \mathbb{N}$, m^n is well-defined.* □

Proof. Fix m and apply Corollary 30 with $A = \mathbb{N}$, $a = m$ and $g : n \to n \times m$. □

Proposition 51 (Laws of Exponents). *Let $m, n, r \in \mathbb{N}$. Then*

(1) $1^n = 1$,
(2) $m^{n+r} = m^n \cdot m^r$,
(3) $(m^n)^r = m^{nr}$,
(4) $(mn)^r = m^r \cdot n^r$.

□

Proposition 52. *Let $m, n, p \in \mathbb{N}$. Then $m \leq n \Rightarrow m^p \leq n^p$ and $p^m \leq p^n$.* □

Proposition 53. *Let $m, n, p \in \mathbb{N}$. Then $n^p = m^p \Rightarrow n = m$.* □

Proposition 54. *Let $m, n, p \in \mathbb{N}$. If $p > 1$, then $p^n = p^m \Rightarrow n = m$.* □

At this stage we have defined all the basic binary operations on natural numbers.

Exercise 2.4.5.1. Show that $n < 2^n$, for each $n \in \mathbb{N}$.

2.5 Sequences

2.5.1 The Concept

We now define the term *sequence*, and operations on sequences of sets.

Definitions 2.11. *An* infinite sequence *is a function whose domain is* $\mathbb{N}$.

A finite sequence *is a function whose domain is* $s(n)^- = \{1, \ldots, n\}$ *for some* $n \in \mathbb{N}$.

We adopt the convention that if the adjective is omitted, the sequence shall always be taken to be infinite.

2.5.2 Notation

If a is a sequence, we denote $a(n)$ by a_n, as a rule. If a is an infinite sequence, it is also denoted by $(a_n)_{n=1}^\infty$, or $(a_1, a_2, \ldots)$, or sometimes just (a_n). It is also common to abuse the notation, and refer to "the sequence a_n". This is an abuse, because strictly speaking a_n is a value (depending on n), not the function.

For finite sequences a, the corresponding notation is $(a_m)_{m=1}^n$, or $(a_1, \ldots, a_n)$. We refer to such a sequence as an ordered *n–tuple*, or an *n-term sequence*.

If $(a_m)_{m=1}^n$ is a finite sequence, we denote $\{a_m : 1 \leq m \leq n\}$ by $\{a_1, \ldots, a_n\}$. Similarly, if B is some class, then $a_1,\ldots,a_n \in B$ stands for $\{a_m : 1 \leq m \leq n\} \subset B$.

2.5.3 Operations on Sequences of Sets

A sequence of sets is a sequence with values in $\mathcal{S}$.
If $A : \mathbb{N} \to \mathcal{S}$, then we write

$$\bigcup_{n=1}^\infty A_n := \bigcup \mathrm{im} A = \{a : \exists n \in \mathbb{N} \text{ such that } a \in A_n\}.$$
$$\bigcap_{n=1}^\infty A_n := \bigcap \mathrm{im} A = \{a : a \in A_n, \forall n \in \mathbb{N}\}.$$

Similarly, if $A : \{1, \ldots, n\} \to \mathcal{S}$, then

$$A_1 \cup \cdots \cup A_n := \bigcup_{m=1}^n A_m := \bigcup \mathrm{im} A.$$

$$A_1 \cap \cdots \cap A_n := \bigcap_{n=1}^\infty A_n := \bigcap \mathrm{im} A.$$

2.6. MAXIMUM AND MINIMUM

Exercise 2.5.3.1. Show
$$\bigcup_{j=1}^{n+1} A_j = \left(\bigcup_{j=1}^{n} A_j\right) \cup A_{n+1}.$$

$$\bigcap_{j=1}^{n+1} A_j = \left(\bigcap_{j=1}^{n} A_j\right) \cap A_{n+1}.$$

Exercise 2.5.3.2. † How do we know that the image of each sequence is a set?

Exercise 2.5.3.3. † Show that there is a set A such that $\emptyset \in A$ and
$$x \in A \Rightarrow x \cup \{x\} \in A.$$

2.6 Maximum and Minimum

2.6.1 Finite Sets

We know that two elements of a totally-ordered set have a maximum and minimum. We now extend this idea to finite sets of elements:

Proposition 55. *Let $\preceq$ be a total order on a set A, let $n \in \mathbb{N}$, and let $a_1, \ldots, a_n \subset A$. Then the set $B = \{a_1, \ldots, a_n\}$ has a maximum and minimum with respect to $\preceq$.*

Proof. Induction on n. □

2.6.2 Notation

With B as in the statement, we denote $\max B$ by each of the notations
$$\max\{a_1, \ldots, a_n\} = \max_{1 \leq j \leq n} a_j,$$

and use similar notation for the minimum. The total order in question is usually clear from the context; if not it will be indicated explicitly in the text.

Exercise 2.6.2.1. † At a dance each boy dances with at least one girl and each girl dances with at least one boy. No boy dances with all the girls. No girl dances with all the boys. Prove that there are boys x and y and girls s and t such that x dances with s but not with t, and y dances with t but not with s.

2.6.3 Remark

The sequence of operations $m+n$, $m \times n$, m^n may be continued. The Tarski-Ackermann *fourth operation* is defined inductively by the rule:

(1) $\phi_4(m, 1) := m$,
(2) $\phi_4(m, p+1) := m^{\phi_4(m,p)}$.

More generally, the Tarski-Ackermann $(s+1)$-st operation is defined by:

(1) $\phi_{s+1}(m, 1) := m$,
(2) $\phi_{s+1}(m, p+1) := \phi_s(m, \phi_{s+1}(m, p))$.

You will observe that these operations are probably not of any practical importance, since the values grow so rapidly. They are of considerable theoretical interest in the theory of computability. For instance, sequences such as $\{\phi_n(2, n)\}$ (a variant of *Ackermann's sequence*) is one of the most impressively growing sequences ever conceived. The function defined on pairs of nonnegative integers by

$$\begin{aligned} A(0, m) &:= m+1, \\ A(n+1, 0) &:= A(n, 1), \\ A(n+1, m+1) &:= A(n, A(n+1, m)) \end{aligned}$$

is known as Ackermann's function, and the sequence $(A(n,n))$ is Ackermann's sequence.

Exercise 2.6.3.1. † Work out (or at least think about) $\phi_4(2, 5)$ and $\phi_5(2, 5)$.

2.7 Project: Another Approach

My colleague Stefan Bechtluft-Sachs suggested an alternative to the approach taken in this chapter to the definition of the arithmetic operations. The student might like to try to fill in the details of the following outline, as a project:

1. Define the *n-th iterate* f^n of a function $f : \mathbb{N} \to \mathbb{N}$ inductively, as follows. Let $f^1 := f$, and then $f^{s(n)}(m) := f(f^n(m))$ for all $n \in \mathbb{N}$ and all $m \in \mathbb{N}$. Here, for each fixed m one uses Corollary 30 with g replaced by f, and the function 'f' defined is $n \mapsto f^n(m)$.

2. Then define $m + n := s^m(n)$ for all $m, n \in \mathbb{N}$, and prove that $+$ has all the same properties as before.

3. Define the function '$n+$' by setting $n + (m) := n + m$ for all $m \in \mathbb{N}$.

4. Then define $1 \times n := n$ and $s(m) \times n = (n+)^m + n$ for all $m, n \in \mathbb{N}$, and prove $m \times n$ has all the same properties as before.

5. Finally, figure out how to define n^m in terms of iteration of the function $n \times : \mathbb{N} \to \mathbb{N}$, defined by $n \times (m) := n \times m$.

3: The Integers

The theory of the integers formalises our intuitive notion of counting whole numbers of things, including debts and nothing.

3.1 Axioms

3.1.1 We assume:

Axiom 3.1 (Integers). *We take as given a set $\mathbb{Z}$ of atoms, called integers, containing $\mathbb{N}$. We assume that there is a distinguished atom $0 \in \mathbb{Z}$, and that there is an injective function $- : \mathbb{N} \to \mathbb{Z}$.*

Axiom 3.2 (Trichotomy). *If $m \in \mathbb{Z}$, then exactly one of these is true: (1) $m = 0$, (2) $m \in \mathbb{N}$, or (3) $\exists! n \in \mathbb{N}$ with $m = -(n)$.*

This implies that 0 is not a natural number. We could rephrase the axiom as saying that

$$\mathbb{Z} = \mathbb{N} \mathbin{\dot\cup} \{0\} \mathbin{\dot\cup} \mathrm{im} -.$$

3.1. AXIOMS

Definition 3.1. *We define* $-n := -(n)$ *for all* $n \in \mathbb{N}$. *We define* $-(-n)$ *to be* n, *for* $n \in \mathbb{N}$, *and* -0 *to be* 0.

We also refer to natural numbers as *positive integers*. An integer m with $-m \in \mathbb{N}$ is called a *negative integer*. We also use the terms *nonpositive* and *nonnegative* in the obvious senses.

Proposition 56. *We have* $-(-m) = m$ *for all* $m \in \mathbb{Z}$.

Proof. Consider the three cases $m = 0$, $m \in \mathbb{N}$ and $-m \in \mathbb{N}$. □

Proposition 57. *The function* $n \mapsto -n$ *is a bijection of* $\mathbb{Z}$ *onto* $\mathbb{Z}$.

Proof. It is a well-defined function by Axiom 3.2 and Definition 3.1. It is injective by the last proposition, and it is surjective by the same proposition. (Cf. Exercise 1.5.6.7 on page 23.)

Definition 3.2. *We extend the total order* $\leq$ *on* $\mathbb{N}$ *to a relation (also denoted* $\leq$*) on* $\mathbb{Z}$, *by insisting that*

1. *If* $x \in \mathbb{N}$, *then* $-x \leq 0 \leq x$.

2. *If* $x, y \in \mathbb{N}$ *and* $x \leq y$, *then* $-y \leq -x$.

We define $x < y$, *as before, to mean that* $x \leq y$ *and* $x \neq y$. *Similarly, we define* $x \geq y$ *and* $x > y$ *for pairs of integers.*

Proposition 58. $\mathbb{Z}$ *is totally-ordered by* $\leq$. □

Next, we extend the binary arithmetic operations of addition, subtraction, and multiplication from $\mathbb{N}^2$ to $\mathbb{Z}^2$, as follows:

3.1.2 Defining +

For $x \in \mathbb{Z}$ we define $x + 0 := 0 + x := x$.

For $x, y \in \mathbb{N}$ we define $x + (-y) = (-y) + x$ to be

1. $x - y$, if $y < x$,

2. 0, if $y = x$, and

3. $-(y - x)$, if $y > x$.

and we define
$$(-x) + (-y) := -(x + y).$$

3.1.3 Defining −

For $x, y \in \mathbb{Z}$, we define

$$x - y := x + (-y).$$

This is consistent with the previous definition given for the case $1 \leq y < x$, because in that case $x + (-y)$ is, by definition (case 1), $x - y$.

3.1.4 Defining ×

For $x \in \mathbb{Z}$ we define $0 \times x := x \times 0 := 0$.
  For $x, y \in \mathbb{N}$ we define

$$x \times (-y) := (-y) \times x = -(x \times y),$$

and

$$(-x) \times (-y) := x \times y.$$

3.1.5 Precedence

We use the usual BODMAS rule about the precedence of × over + or −, i.e. $a \times b \pm c$ denotes $(a \times b) \pm c$ and $a \pm b \times c$ denotes $a \pm (b \times c)$. These operations have higher precedence than the relations $<, \leq, >, \geq$ and $=$. In expressions involving sets, the operations $\cup, \cap$ and $\sim$ have equal precedence, higher than $\bigcup_{i=1}^{n}$ and $\bigcap_{i=1}^{n}$, then $\subset$ and $=$. The logical operators all have lower preference than the arithmetic or set operators. The negation $\neg$ comes highest, then the operators $\vee$ and $\wedge$, then $\forall$ and $\exists$, then $\Rightarrow$ and $\Leftrightarrow$.

Exercise 3.1.5.1. † Put brackets into these expressions as dictated by the precedence rules:

1. $a + b \times c + d < a - f \Rightarrow f \times g = 0 \wedge h < 0$.

2. $a \in A \cup B \vee b \in C \cap D \Leftrightarrow c \in D \cap \bigcup_n E_n \cup F_n$.

3.2. POWERS

3.1.6 Properties of <, +, − and ×

This order and these operations have the following properties:

Proposition 59. *Let $x, y, z \in \mathbb{Z}$. Then:*
(1) If $x < y$, then $x + z < y + z$.
(2) $x + (y + z) = (x + y) + z$.
(3) $x + y = y + x$.
(4) $x + 0 = x$.
(5) $x + (-x) = 0$.
(6) $x + y = x + z \Rightarrow y = z$.
(7) $-(x + y) = (-x) + (-y)$.
(8) $x - y = x - z \Rightarrow y = z$. □

Proposition 60. *Let $x, y, z \in \mathbb{Z}$. Then:*
(1) If $x < y$ and $z > 0$, then $x \times z < y \times z$.
(2) $x \times (y \times z) = (x \times y) \times z$.
(3) $x \times y = y \times x$.
(4) $x \times 1 = x$.
(5) $(-1) \times x = -x$.
(6) If $x \neq 0$ and $x \times y = x \times z$, then $y = z$.
(7) If $x \times y = 0$, then $x = 0$ or $y = 0$.
(8) $-(x \times y) = (-x) \times y$.
(9) $x \times (y + z) = x \times y + x \times z$. □

3.1.7 The Notation $\mathbb{Z}_+$

We denote the set of nonnegative integers $\mathbb{N} \cup \{0\}$ by $\mathbb{Z}_+$.

3.1.8 Factorial 0

We extend the factorial function from $\mathbb{N}$ to $\mathbb{Z}_+$ by defining $0! := 1$. We do not define the factorial of a negative integer.

3.2 Powers

3.2.1 Powers of Negative Integers

We have previously defined a^b for natural numbers a and b. We extend this definition somewhat:

Definition 3.3. For $x, y \in \mathbb{N}$, we define $(-x)^y := x^y$ if y is even, and $(-x)^y := -x^y$ if y is odd.
We define
$$0^y := 0, \ \forall y \in \mathbb{N},$$
and
$$x^0 := 1, \ \forall x \in \mathbb{Z}, x \neq 0.$$

We shall not define 0^0 at all. We are not yet in a position to define a^b for negative integers b. So far, the domain of $(x, y) \mapsto x^y$ is
$$(\mathbb{Z} \times \mathbb{N}) \cup (\{x \in \mathbb{Z} : x \neq 0\} \times \{0\}).$$

With this extended definition, the laws of exponents continue to hold, whenever both sides make sense.

Proposition 61. Let $0 \neq x \in \mathbb{Z}$, $y, z \in \mathbb{Z}_+$. Then
(1) $x^{(y+z)} = x^y \times x^z$.
(2) $(x^y)^z = x^{y \times z}$.
(3) $(x \times y)^z = (x^z) \times (y^z)$. □

Exercise 3.2.1.1. † Consider what becomes of this proposition if we drop the assumption $x \neq 0$.

Proposition 62. For each $n \in \mathbb{Z}_+$, we have $n < 2^n$.

Proof. Induction. □

Suppose $x, y \in \mathbb{Z}$. Prove:

Exercise 3.2.1.2. † If $x + y = x$, then $y = 0$ ('0 is unique').

Exercise 3.2.1.3. If $x + y = 0$, then $y = -x$.

Exercise 3.2.1.4. $-(-x) = x$.

Exercise 3.2.1.5. $-(x - y) = y - x$.

3.3 Characteristic Functions

3.3.1 The Function χ_A

Definition 3.4. *If A is a subset of a set X, then the* characteristic function *of the set A with respect to X is the function $\chi_A : X \to \{0, 1\}$ defined by*

$$\chi_A(x) := \begin{cases} 1 & , \quad \text{if } x \in A, \\ 0 & , \quad \text{if } x \in X \sim A. \end{cases}$$

In other words, for $x \in X$ we have the equivalences

$$x \in A \iff \chi_A(x) = 1,$$

and

$$x \in X \sim A \iff \chi_A(x) = 0.$$

For a fixed set X, the map $A \mapsto \chi_A$ is a bijection between the power set 2^X and the set of all functions from X to $\{0,1\}$, i.e. it is a bijection between 2^X and $\{0,1\}^X$. This is the historical reason for the notation 2^X. In one popular theory about the foundations of mathematics, the numbers were constructed from primitive objects of set theory, by defining $0 := \emptyset$, $1 := \{0\}$, $2 := \{0, 1\}$, $3 := \{0, 1, 2\}$, and so on. (This is not what we do in this book, and it can only be done if additional axioms are added to the set theoretic axioms we used.) On the basis of this construction, 2^X and $\{0,1\}^X$ denoted *one and the same thing*, namely the set of characteristic functions of subsets of X. In view of the bijective correspondence between characteristic functions and subsets, it was considered reasonable to use the same notation for both the set of characteristic functions and the set of subsets.

To be clear about it, our 2 is an atom, and is different from $\{0, 1\}$. Our 2 is *not* a set, so for sets X the notation 2^X cannot be read as "the set of all functions from X into 2". It just denotes the power set.

4: The Rational Numbers

The theory of rational numbers formalises our intuitive notions of calculating with fractional parts of things.

We now set up the system $\mathbb{Q}$ of rational numbers, equipped with its unary operations of minus and reciprocal and its binary arithmetic operations of addition, subtraction, multiplication and exponentiation. The key properties of the rationals are summarised in Section 4.3[1]. Having read them, the reader who is willing to take them for granted can then skip to Section 4.4.

4.1 Axioms and Arithmetical Properties

4.1.1 Axioms for $\mathbb{Q}$

Axiom 4.1 (Rationals). *We take as given a set $\mathbb{Q}$ of atoms, called rational numbers, containing the set $\mathbb{Z}$ of integers, and equipped with a binary operation $\times$, extending the operation $\times$ already defined on $\mathbb{Z}$, and such that*

1. *(Associative)* $(x \times y) \times z = x \times (y \times z)$, *for all* $x, y, z \in \mathbb{Q}$.

2. *(Commutative)* $x \times y = y \times x$, *for all* $x, y \in \mathbb{Q}$.

3. *(Identity)* $x \times 1 = x$, *for all* $x \in \mathbb{Q}$.

---

[1] Professionals summarise these properties by saying that the rational numbers, together with their basic arithmetic operations, form a minimal ordered field. They can also be described as forming a minimal field of characteristic zero.

4.1. AXIOMS AND ARITHMETICAL PROPERTIES

> 4. *(Inverse or Reciprocal)* For each nonzero $x \in \mathbb{Q}$, there exists a $y \in \mathbb{Q}$ such that $x \times y = 1$.
>
> 5. *(Fraction)* For each $x \in \mathbb{Q}$, there exists $y \in \mathbb{N}$ such that $x \times y \in \mathbb{Z}$.

The idea behind this axiom is that $\mathbb{Q}$ is *just* big enough to allow us to solve $a \times x = b$ for the unknown x, whenever $a \in \mathbb{N}$ and $b \in \mathbb{Z}$.

The 'associative' part of the axiom allows us to write $x \times y \times z$ without ambiguity. As before, we also denote $x \times y$ by $x \cdot y$ and by xy.

4.1.2 Notation

We let $\mathbb{Q}^{\times} := \{x \in \mathbb{Q} : x \neq 0\}$.

4.1.3 Reciprocal

Lemma 63. *Let $x, y \in \mathbb{Q}$. If $xy = 1$, then $yx = 1$.*

Proof. Obvious from commutativity. □

Lemma 64. *For each $x \in \mathbb{Q}^{\times}$ there exists a* unique *$y \in \mathbb{Q}$ such that $xy = 1$.*

Proof. The existence follows from the 'inverse' part of the axiom. To see uniqueness, suppose $xy = xz = 1$. Then $yx = 1$, so
$$y = y1 = yxz = 1z = z.$$
□

4.1.4 Notation

We denote the unique y with $xy = 1$ by x^{-1}. We call x^{-1} the *reciprocal*[2] of x. Note that $1^{-1} = 1$, because $1 \times 1 = 1$.

Proposition 65. *Let $x \in \mathbb{Q}^\times$. Then $x^{-1} \neq 0$.*

Proof. Suppose $x0 = 1$. Choose $y \in \mathbb{N}$ such that $xy \in \mathbb{Z}$. Then $y = y1 = (xy)0 = 0$, a contradiction. □

Proposition 66 (Cancellation Law). *Let $x, y, z \in \mathbb{Q}$, $x \neq 0$, and $xy = xz$. Then $y = z$.*

Proof.
$$y = 1y = x^{-1}xy = x^{-1}xz = 1z = z.$$
□

4.1.5 Division

Definition 4.1. *For $x \in \mathbb{Q}$ and $y \in \mathbb{Q}^\times$, we define*
$$\frac{x}{y} := x \times y^{-1}.$$

We also define
$$x/y := \frac{x}{y}.$$

Proposition 67. *Let $x, y, z \in \mathbb{Q}$ and $a, b, c \in \mathbb{Q}^\times$. Then the following hold:*
(1) $x/1 = x$.
(2) $1/a = a^{-1}$.
(3)
$$\frac{x}{a} = \frac{y}{b} \Leftrightarrow xb = ya.$$

(4) $(ab)^{-1} = a^{-1}b^{-1}$.

---

[2] It is also called the inverse of x. This is not to be confused with the inverse of a function or relation. Rational numbers are not functions.

4.1. AXIOMS AND ARITHMETICAL PROPERTIES

(5) $(xa)/(ba) = x/b.$
(6)
$$\frac{x}{a} \times \frac{y}{b} = \frac{xy}{ab}.$$

(7)
$$\frac{\left(\dfrac{x}{a}\right)}{\left(\dfrac{b}{c}\right)} = \frac{xc}{ab}.$$

(8)
$$\frac{x}{a} = \frac{y}{a} \Rightarrow x = y.$$

(9)
$$\left(\frac{a}{b}\right)^{-1} = \frac{b}{a}.$$

Proof. (1) $x/1 = x1^{-1} = x1 = x.$
(2) $1/a = 1a^{-1} = a^{-1}.$
(3) If $x/a = y/b$, then $xa^{-1} = yb^{-1}$, so $xb = xb1 = xbaa^{-1} = xa^{-1}ba = yb^{-1}ba = ya$. In the other direction, if $xb = ya$, then $xa^{-1} = x1a^{-1} = xbb^{-1}a^{-1} = yaa^{-1}b^{-1} = y1b^{-1}.$
(4) $(ab)(a^{-1}b^{-1}) = aa^{-1}bb^{-1} = 1 \cdot 1 = 1.$
(5) $(xa)/(ba) = xa(ba)^{-1} = xaa^{-1}b^{-1} = x1b^{-1} = x/b.$
(6)-(9): Exercise. □

4.1.6 Numerators and Denominators

Definition 4.2. *Let $x \in \mathbb{Q}$, then we define*

$$\mathrm{denom}(x) := \min\{y \in \mathbb{N} : xy \in \mathbb{Z}\},$$

and

$$\mathrm{num}(x) := x \times \mathrm{denom}(x).$$

This definition makes sense, because the 'fraction' part of the Axiom of Rationals tells us that the set $\{y \in \mathbb{N} : xy \in \mathbb{Z}\}$ is nonempty, and because $\mathbb{N}$ is well-ordered.

Proposition 68. *Let $x \in \mathbb{Q}$. Then:*
(1) $\operatorname{denom}(x) \in \mathbb{N}$ and $\operatorname{num}(x) \in \mathbb{Z}$.
(2) $x \in \mathbb{Z} \Leftrightarrow \operatorname{denom}(x) = 1 \Leftrightarrow \operatorname{num}(x) = x$.
(3)
$$x = \frac{\operatorname{num}(x)}{\operatorname{denom}(x)}.$$

(4) The map $x \mapsto (\operatorname{num}(x), \operatorname{denom}(x))$ is an injection from $\mathbb{Q}^\times$ into
$$\{(n, m) : n \in \mathbb{Z}, m \in \mathbb{N}\}.$$

(5) The mapping
$$\begin{cases} \mathbb{Z} \times \mathbb{N} & \to \quad \mathbb{Q} \\ (n, m) & \mapsto \quad \frac{n}{m} \end{cases}$$
is surjective. □

When we write a rational number x in the form n/m, with $n \in \mathbb{Z}$ and $m \in \mathbb{N}$, we are said to represent it as a *fraction*. The particular form $\operatorname{num}(x)/\operatorname{denom}(x)$ is called its representation in *lowest terms*, or its *canonical form*.

Lemma 69. $x \times 0 = 0$ *for each $x \in \mathbb{Q}$.*

Proof. Otherwise, take $y = (x0)^{-1}$. Then $y \neq 0$ and $yx0 = 1$.
Let $a = \operatorname{denom}(x)$, $b = \operatorname{denom}(y)$. Then $a, b \in \mathbb{N}$, and
$$ab = ab1 = abxy0 = (ax)(by)0 = 0.$$
But $ab \in \mathbb{N}$ so this is impossible. □

4.1.7 Unary Minus

Lemma 70. *Let $x, y \in \mathbb{Z}$ and $a, b \in \mathbb{N}$. Then*
$$\frac{x}{a} = \frac{y}{b} \Rightarrow \frac{(-x)}{a} = \frac{(-y)}{b}.$$

4.1. AXIOMS AND ARITHMETICAL PROPERTIES

Proof. We get $xb = ya$, so by Proposition 59

$$(-x)b = -(xb) = -(ya) = (-y)a,$$

and this yields the result. □

This allows us to define the *negative* of a rational unambiguously as follows:

Definition 4.3. *We define*

$$-\left(\frac{x}{a}\right) := \frac{-x}{a},$$

whenever $x \in \mathbb{Z}$ *and* $a \in \mathbb{N}$.

Proposition 71. *let* $x, y \in \mathbb{Q}$ *and* $a, b \in \mathbb{Q}^\times$. *Then:*
(1) $\mathrm{num}(-x) = -\mathrm{num}(x)$.
(2) $\mathrm{denom}(-x) = \mathrm{denom}(x)$.
(3) $(-1)x = -x$.
(4) $x(-y) = -(xy)$.
(5) $(-1)(-1) = 1$, *so* $(-1)^{-1} = -1$.
(6) $(-a)^{-1} = -(a^{-1})$
(7)
$$\frac{-a}{b} = \frac{a}{-b} = -\frac{a}{b}.$$

□

4.1.8 Addition

Lemma 72. *Let* $x, y, x', y' \in \mathbb{Z}$ *and* $a, b, a', b' \in \mathbb{N}$. *Then*

$$\left.\begin{matrix}\frac{x}{a} = \frac{x'}{a'}\\ \frac{y}{b} = \frac{y'}{b'}\end{matrix}\right\} \Rightarrow \frac{xb + ya}{ab} = \frac{x'b' + y'a'}{a'b'}.$$

Proof. We have $xa' = x'a$ and $yb' = y'b$, so $(xb + ya)a'b' = xba'b' + yaa'b' = x'abb' + y'baa' = ab(x'b' + y'a')$. This gives the result. □

This lemma allows us to define addition on $\mathbb{Q}$ unambiguously as follows:

Definition 4.4. *For $x, y \in \mathbb{Z}$ and $a, b \in \mathbb{N}$, we define*

$$\left(\frac{x}{a}\right) + \left(\frac{y}{b}\right) := \frac{xb + ya}{ab}.$$

Since $x = x/1$ when $x \in \mathbb{Z}$, this definition agrees with that already in place for integers.

Exercise 4.1.8.1. If $x, y \in \mathbb{Q}$ and $a, b \in \mathbb{Q}^\times$, then show

$$\left(\frac{x}{a}\right) + \left(\frac{y}{b}\right) = \frac{xb + ya}{ab}.$$

Proposition 73. *Let $x, y, z \in \mathbb{Q}$. Then:*
(1) (Associative) $x + (y + z) + (x + y) + z$.
(2) (Commutative) $x + y = y + x$.
(3) (Zero) $x + 0 = x$.
(4) (Reciprocal) $x + (-x) = 0$
(5) ($\times$ distributes over $+$) $x(y + z) = xy + xz$.

Proof. We prove the distributive law, and leave the rest as an exercise.

Let $x = u/a$, $y = v/b$ and $z = w/c$, with $u, v, w \in \mathbb{Z}$ and $a, b, c \in \mathbb{N}$. Then a calculation gives

$$x(y+z) = \frac{uvc + uwb}{abc} = \frac{auvc + auwb}{a^2bc} = xy + xz.$$

$\square$

4.1.9 Subtraction

Definition 4.5. *We define $x - y := x + (-y)$ whenever $x, y \in \mathbb{Q}$.*

Proposition 74. *Let $x, y, z \in \mathbb{Q}$. Then:*
(1) $x - x = 0$.
(2) $x - 0 = x$.
(3) $0 - x = -x$.

4.1. AXIOMS AND ARITHMETICAL PROPERTIES

(4) $-(x+y) = (-x) + (-y)$.
(5) $x - (y+z) = (x-y) + z$
(6) $x - (y-z) = (x-y) + z = x + (z-y)$
(7) $x \times (y-z) = (x \times y) - (x \times z)$. □

We use the (standard) convention that precedence in an unbracketed expression involving only $+$ and $-$ operations, such as $a - b + c - d - e + f$ is from left to right. In other words, an operation to the left is performed first. Thus

$$a - b + c - d - e + f := ((((a-b) + c) - d) - e) + f.$$

4.1.10 Order

We know that $\text{num}(x) = x$ when $x \in \mathbb{Z}$, so we can extend the definition of $>$ from $\mathbb{Z}$ to $\mathbb{Q}$ as follows:

Definition 4.6. *Given $x \in \mathbb{Q}$, we say that (1) $x > 0$ if $\text{num}(x) > 0$, and (2) $x < 0$ if $\text{num}(x) < 0$.*

For $x, y \in \mathbb{Q}$, we say that $x < y$ (and also that $y > x$) if $y - x > 0$. We define $x \leq y$ and $x \geq y$ as usual.

We extend the use of the terms *positive*, *negative*, *nonpositive* and *nonnegative* to rationals, as usual.

Proposition 75. *Let $x, y, z \in \mathbb{Q}$. Then:*
(1) $x > 0$ if and only if $-x < 0$.
(2) $x > y \Rightarrow x + z > y + z$.
(3) $x > y \Rightarrow x - z > y - z$.
(4) If $x > y$ and $z > 0$, then $xz > yz$.
(5) If $x > y$ and $z < 0$, then $xz < yz$.
(6) If $x > y > 0$ or $0 > x > y$, then $1/y > 1/x$.
(7) If $x > 0$, then $1/x > 0$.
(8) If $x < 0$, then $1/x < 0$.
(9) If $x > y$ and $z > 0$, then

$$\frac{x}{z} > \frac{y}{z}.$$

□

4.2 Powers

4.2.1 Integral Exponents

We now extend the definition of a^b to certain rational a and b. The definitions are forced if we want to preserve the laws of exponents.

Definition 4.7. *For $a \in \mathbb{Q}$, we define a^n inductively for $n \in \mathbb{N}$ by*
$$\begin{aligned} a^1 &:= a, \\ a^{n+1} &:= a \times a^n. \end{aligned}$$
We define $a^0 := 1$, for each nonzero $a \in \mathbb{Q}$.
For $a \in \mathbb{Q}$, $a \neq 0$, and $n \in \mathbb{N}$, we define
$$a^{-n} := \frac{1}{a^n}.$$

This is consistent with the previous definition for $a \in \mathbb{Z}$.

Exercise 4.2.1.1. † Show $a^{-n} = (a^{-1})^n$, for all $a \in \mathbb{Q}^\times$ and $n \in \mathbb{N}$.

Proposition 76. *Let $a, b \in \mathbb{Q}$ and $m, n \in \mathbb{N}$.*
If $0 < a < b$, then $0 < a^n < b^n$.
If $a > 1$ and $m < n$, then $1 < a^m < a^n$.
If $0 < a < 1$ and $m < n$, then $1 > a^m > a^n > 0$.

Proof. Use induction. □

4.2.2 Why not Rational Exponents?

We are not yet in a position to define nonintegral powers of general rationals:

Theorem 77. *There is no rational number r with $r^2 = 2$.*

Proof. Suppose that there exist $m \in \mathbb{Z}$ and $n \in \mathbb{N}$ with
$$\left(\frac{m}{n}\right)^2 = 2.$$

Cancelling any common factor 2, we may suppose that at least one of m and n is odd. Now $m^2 = 2n^2$, hence m^2 is even, hence m is even. Write $m = 2p$, for some $p \in \mathbb{Z}$. Then $2p^2 = n^2$, hence n^2 is even, so n is also even. This is impossible. Thus no such m, n exist. □

4.3 Key Properties of $\mathbb{Q}$

This is a summary.

The set $\mathbb{Q} = \mathbb{Q}^\times \dot\cup \{0\}$ of rational numbers contains the integers, and has binary operations $+$ and $\times$, and a unary operation $-$, extending the corresponding operations on $\mathbb{Z}$, a unary operation $x \mapsto x^{-1}$ on $\mathbb{Q}^\times$, and a total order $\leq$, extending the order on $\mathbb{Z}$, and such that for all $x, y, z \in \mathbb{Q}$:
(1) $x + (y + z) + (x + y) + z$.
(2) $x + y = y + x$.
(3) $x + 0 = x$.
(4) $x + (-x) = 0$
(5) $(x \times y) \times z = x \times (y \times z)$.
(6) $x \times y = y \times x$.
(7) $x \times 1 = x$.
(8) If $x \neq 0$, then $x \times x^{-1} = 1$.
(9) There exists $n \in \mathbb{N}$ such that $x \times n \in \mathbb{Z}$.
(10) $x \times (y + z) = x \times y + x \times z$.
(11) Exactly one of $x > 0$, $x = 0$, and $0 > x$ is true.
(12) $x > y \Rightarrow x + z > y + z$.
(13) If $x > y$ and $z > 0$, then $x \times z > y \times z$.

4.4 Binomial Coefficients

Recall the factorial $n!$, which was defined in Subsection 3.1.8 for $n \in \mathbb{Z}_+$.

4.4.1 The Concept

Definition 4.8. *We define the* binomial coefficient
$$\binom{n}{m} := \frac{n!}{m!(n-m)!},$$

for integers n, m with $0 \leq m \leq n$. We define
$$\binom{n}{m} := 0,$$
for integers n, m with $0 \leq n < m$.

Theorem 78 (Pascal's Triangle). Let $n, m \in \mathbb{Z}_+$. Then
$$\binom{n+1}{m+1} = \binom{n}{m+1} + \binom{n}{m}.$$

Proof. When $m > n$, both sides are 0, and when $m = n$, both sides are 1, so it remains to prove the case $m < n$. This case does not occur when $n = 0$, so the result holds for $n = 0$ and all $m \in \mathbb{Z}_+$. We proceed by induction on n.

Assume that the formula holds for a given $n \in \mathbb{Z}^+$ and for all $m \in \mathbb{Z}_+$. Then for $m < n$, we have

$$\begin{aligned}
\binom{n}{m+1} + \binom{n}{m} &= \frac{n!}{(m+1)!(n-m-1)!} + \frac{n!}{m!(n-m)!} \\
&= \frac{n!}{m!(n-m-1)!}\left\{\frac{1}{m+1} + \frac{1}{n-m}\right\} \\
&= \frac{n!}{m!(n-m-1)!}\left\{\frac{n+1}{(m+1)(n-m)}\right\} \\
&= \binom{n+1}{m+1}.
\end{aligned}$$

So the result holds for $n + 1$. □

Corollary 79. Let $n, m \in \mathbb{Z}$, $0 \leq m \leq n$. Then $\binom{n}{m}$ is a nonnegative integer. □

Proof. Induction on n. □

Exercise 4.4.1.1. If $r, s \in \mathbb{Q}$ are nonzero, $0 < r < s$, and
$$s - r < \frac{1}{\text{denom}(r)\text{denom}(s)},$$
show that $r = s$.

4.4. BINOMIAL COEFFICIENTS

Exercise 4.4.1.2. † Each positive $r \in \mathbb{Q}$ has a unique expression as a (terminating) simple continued fraction

$$a_1 + \cfrac{1}{a_2+} \cfrac{1}{a_3 + \cdots} \cfrac{1}{a_n},$$

where the a_j are integers, $a_1 \geq 0$, $a_j > 0$ when $j > 1$, and $a_n > 1$.

Exercise 4.4.1.3. † Each positive $r \in \mathbb{Q}$ has a unique expression

$$a_1 + \frac{a_2}{2!} + \frac{a_3}{3!} + \cdots + \frac{a_k}{k!},$$

where the a_j are integers, $0 \leq a_1$, $0 \leq a_2 < 2$, $0 \leq a_3 < 3,\ldots$, $0 \leq a_k < k$.

5: The Real Numbers

5.1 Overview

The theory of real numbers formalises our intuitive notions about reckoning with quantities that can change in a continuous way. It was invented by people[1]. It is a very sophisticated business.

You will observe that we are steadily adding new kinds of numbers, and this can be viewed as a reaction to the occurrence of equations that don't necessarily have solutions among the numbers so far available. Thus, to solve

$$x + a = b$$

with $a, b \in \mathbb{N}$, we need $\mathbb{Z}$, and to solve

$$ax = b$$

with $a, b \in \mathbb{Z}$, we need $\mathbb{Q}$. So you might expect our next move to be a reaction to the equations (1) $x^a = b$ or (2) $a^x = b$, with $a, b \in \mathbb{Q}$. To deal with (1) we need to add certain so-called 'algebraic' numbers to the system, and to deal with (2) we need to add certain 'logarithms'. However, this would still leave us with more problems, and the process could be endless. So, in fact, our next move is to plunge through a 'wormhole' to a number system that will allow us to solve all these equations at once (provided they can be solved at all). This number system

---

[1] Leopold Kronecker: "God invented the integers; all else is the work of man."

is the system of complex numbers. The wormhole has a way-station, the system of real numbers.[2]

We now set up the real number system for use. Many books on analysis begin by taking as axioms the key properties listed in Section 5.5[3] below.

We follow another route, assuming fewer axioms, and building on our previous axioms on sets, natural numbers, integers and rationals. Our approach follows the 'evolutionary route' of the real number system, which is also the educational route by which school-children approach the 'number line'.

5.2 The Real Number System

5.2.1 Real Numbers are Atoms

Axiom 5.1 (Real Atoms). *We take as given a set $\mathbb{R}$ of atoms, called real numbers, containing the set $\mathbb{Q}$ of rational numbers.*

5.2.2 Order

Axiom 5.2 (Order). *We assume that there is a strict order $>$ on $\mathbb{R}$, extending the strict order $>$ on $\mathbb{Q}$.*

---

[2] It could be argued that we go too far, and that a smaller set of numbers would be enough for all 'reasonable' problems. This position is quite defensible, in fact. There is a sense in which 'most' real numbers are convenient fictions, useful to have around for various theoretical purposes, devoid of any but generic properties, and never appearing in the solution of any practical problem. However, a fuller discussion of these matters would take us far afield.

[3] Professionals summarise these properties by saying that the real numbers, together with their basic arithmetic operations, form a complete ordered field.

Recall that this implies that given $x, y \in \mathbb{R}$, exactly one of the three statements $x = y$, $x > y$ and $y > x$ is true. Also $x > y > z$ implies $x > z$.

We say that x is *positive* if $x > 0$, and x is *negative* if $x < 0$. We write $x < y$ to mean $y < x$, $x \leq y$ to mean $(x = y) \vee (x < y)$ and $x \geq y$ to mean $(x = y) \vee (x > y)$.

To save paper, we will always understand $x > y$ to mean that *both x and y are real numbers*, and $x > y$. Similarly, the occurrence of $>$, $\leq$, or $\geq$ between two names indicates that both names denote real numbers.

We write $x < y < z$ to mean $x < y$ and $y < z$. Similarly for $x \leq y < z \leq w$, and so on.

Let $x, y, z \in \mathbb{R}$. We say that z is *between x and y* if $x < z < y$ or $y < z < x$.

5.2.3 Notation

We denote the set of nonzero real numbers by $\mathbb{R}^{\times}$, and the set of nonnegative real numbers by $\mathbb{R}_+$. [4]

For a and b real, we denote the various kinds of *intervals* as follows:

$$\begin{aligned}
[a, b] &:= \{x \in \mathbb{R} : a \leq x \leq b\}, \\
(a, b) &:= \{x \in \mathbb{R} : a < x < b\}, \\
[a, b) &:= \{x \in \mathbb{R} : a \leq x < b\}, \\
(a, b] &:= \{x \in \mathbb{R} : a < x \leq b\}, \\
[a, +\infty) &:= \{x \in \mathbb{R} : a \leq x\}, \\
(a, +\infty) &:= \{x \in \mathbb{R} : a < x\}, \\
(-\infty, b] &:= \{x \in \mathbb{R} : x \leq b\}, \\
(-\infty, b) &:= \{x \in \mathbb{R} : x < b\}. \\
(-\infty, +\infty) &:= \mathbb{R}
\end{aligned}$$

(The symbols $\pm\infty$ occurring here do not stand for real numbers; the notation is just a formal convenience.)

Intervals of the form $[a, b]$, $[a, b)$, $(a, b]$ or (a, b) with real a and b are called *bounded intervals*. The rest are called *unbounded*. Those of the form (a, b), $(a, +\infty)$, $(-\infty, b)$ or $(-\infty, +\infty)$

---
[4]Some authors use $\mathbb{R}_+$ to denote the set of positive real numbers, instead.

5.2. THE REAL NUMBER SYSTEM

are called *open intervals*. Those of the form $[a, b)$ or $(a, b]$ are called *half-open*. Those of the form $[a, b]$, $[a, +\infty)$, $(-\infty, b]$, or $(-\infty, +\infty)$ are called *closed*[5].
Note that $\mathbb{R}_+ = [0, +\infty)$.

Exercise 5.2.3.1. Classify the following intervals as bounded or unbounded, open or not open, closed or not closed: $(0, 2]$, $(0, +\infty)$, $(-\infty, +\infty)$, $(-3, 3)$.

5.2.4 Remark

The set $\{a\}$ has a as its sole element. The set $\{a, b\}$, has a and b as its only elements. Make sure you understand the difference between the closed interval $[0, 1]$, the open interval $(0, 1)$, the point $(0, 1)$ in $\mathbb{R} \times \mathbb{R}$ (the Cartesian plane[6], of which was shall see much more eventually) and the doubleton $\{0, 1\}$. Note especially the **ambiguity** in the notation (a, b), when a and b are real numbers. This can be resolved only by appeal to context, and it is a fortunate coincidence that much of the time such resolution is possible. This ambiguous usage is deeply engrained in mathematical practice; it can be eliminated by systematically using $]a, b[$ for the open interval and this is done by some authors.

5.2.5 Dense Rationals

We also assume:

---

[5] In real analysis, the terms bounded, unbounded, open, and closed are applied to more general sets than intervals, in a way that is consistent, for intervals, with the usage defined here. So you should not assume, if you dip into an analysis text, that a set described as, for instance, open and bounded, is actually an interval.

[6] $\mathbb{R} \times \mathbb{R}$ is the set
$$\{(x, y) : x, y \in \mathbb{R}\}.$$
This is the "first" Cartesian product, in that it is essentially the one invented by R. Descartes.

> **Axiom 5.3 (Dense Rationals).** *If $a, b \in \mathbb{R}$ and $a < b$, then there exist $q, r, s \in \mathbb{Q}$ with $q < a < r < b < s$.*

This says that no real number is greater than all the rationals, or less than all the rationals, and that there is a rational between any two reals. It thus implies that no real number is greater than all the others, or is less than all the others.

5.2.6 Bounds

Recall from Subsection 1.7.5 the terminology about upper bounds and lower bounds with respect to a partial order on a set. When we are dealing with the set $\mathbb{R}$, we implicity refer to the order $\leq$, which is of course a total order. Let's recap what this means:

Let $A \subset \mathbb{R}$. A number $b \in \mathbb{R}$ is an *upper bound* for the set A if $x \leq b$, $\forall x \in A$. We say that b is a *lower bound* for E if $x \geq b$, $\forall x \in E$.

The set A is *bounded above* if there exists an upper bound for A. The number b is a *least upper bound* or supremum for A if b is an upper bound for A and $b \leq c$ whenever c is an upper bound for A. If a set E has no upper bound, we say that it is *unbounded above*.

The terms *greatest lower bound* or *infimum* and *unbounded below* are defined in a similar way.

For instance, 2, 3.14 and 5 are all upper bounds for $[0, 1]$. The least upper bound of $[0, 1]$ is 1.

Exercise 5.2.6.1. Give an upper bound for the set $[-1, 2]$. Give two more.

5.2.7 sup and inf

Every number is an upper bound and a lower bound for the empty set. Each set has at most one least upper bound and at most one greatest lower bound.

5.2. THE REAL NUMBER SYSTEM

If a set $E \subset \mathbb{R}$ has a least upper bound, we denote it by $\sup E$. If it has a greatest lower bound, we denote it by $\inf E$. If E is unbounded above, we write $\sup E = +\infty$. If E is unbounded below, we write $\inf E = -\infty$. We write $\sup \emptyset = -\infty$, and $\inf \emptyset = +\infty$.

Thus, if we assert that $\sup E = x$, for some real number x, we are implicitly saying also that E is nonempty and bounded above.

Clearly, if b is an upper bound for E and $c > b$, then c is also an upper bound for E.

The symbols $+\infty$ and $-\infty$ do not represent real numbers, but we adopt the convention that $-(+\infty) = -\infty$ and $-(-\infty) = +\infty$, for later convenience.

Proposition 80. *If $x \in \mathbb{R}$, then*

$$\sup\{r \in \mathbb{Q} : r < x\} = x = \inf\{r \in \mathbb{Q} : x < r\}.$$

Proof. Let $A = \{r \in \mathbb{Q} : r < x\}$ and $B = \{r \in \mathbb{Q} : x < r\}$. Evidently, $\sup A \leq x \leq \inf B$, since x is an upper bound for A and a lower bound for B.

Let $y < x$. Then by the Dense Rationals Axiom, there is a rational r with $y < r < x$, hence $r \in A$, so y is not an upper bound for A. Thus x is the least upper bound of A.

Similarly, x is the greatest lower bound of B. $\square$

5.2.8 Completeness

We assume the following:

> **Axiom 5.4 (Completeness).** *Each nonempty set of real numbers that is bounded above has a least upper bound*[7].

---

[7] It would actually be sufficient to assume this for sets of rationals, with a little extra work.

It follows from the Axiom of Completeness that the set of all upper bounds for a set E is an interval of the form $[u, +\infty)$ or is $\emptyset$. In the first case, we have $u = \sup E$.

5.2.9 Basic properties of sup and inf

It takes practice to develop facility in working with sup and inf. The ability to do this is essential if you are to master analysis. Here we spell out reformulations of basic statements about the relationship between a real number s and the sup and inf of a set $A \subset \mathbb{R}$:

Proposition 81.
(1) $s = \sup A$ *means: (i)* $a \leq s$ *for each* $a \in A$, **and** *(ii) for each* $t < s$, $\exists a \in A$ *with* $t < a$.
(2) $s \neq \sup A$ *means: (i)* $s < a$ *for some* $a \in A$, **or** *(ii)* $\exists t < s$ *such that* $a \leq t$, $\forall a \in A$.
(3) $s > \sup A$ *means:* $\exists t < s$ *such that* $a \leq t$, $\forall a \in A$.
(4) $s < \sup A$ *means:* $\exists a \in A$ *such that* $s < a$.
(5) $s = \inf A$ *means: (i)* $s \leq a$, $\forall a \in A$, **and** *(ii) for each* $t > s$, $\exists a \in A$ *with* $a < t$.
(6) $s \neq \inf A$ *means: (i)* $s > a$ *for some* $a \in A$, **or** *(ii)* $\exists t > s$ *such that* $t \leq a$, $\forall a \in A$.
(7) $s > \inf A$ *means:* $\exists a \in A$ *with* $a < s$.
(8) $s < \inf A$ *means:* $\exists t > s$ *with* $t \leq a$, $\forall a \in A$. □

Exercise 5.2.9.1. Show that for each $x \in \mathbb{R}$,

$$\begin{aligned} x &= \sup\{r \in \mathbb{Q} : r < x\} \\ &= \sup\{r \in \mathbb{Q} : r \leq x\} \\ &= \inf\{r \in \mathbb{Q} : r > x\} \\ &= \inf\{r \in \mathbb{Q} : r > x\}. \end{aligned}$$

Exercise 5.2.9.2. † For each of the following sets, state whether it is bounded above, bounded below, both, or neither:
(a) $\mathbb{Z}$, (b) $\mathbb{N}$, (c) $\{\dfrac{n}{n+1} : n \in \mathbb{N}\}$, (d) $\{x^2(1-x)^2 : x \in \mathbb{Q}\}$.

Exercise 5.2.9.3. † For each of the sets in 5.2.9.2, find its supremum and infimum.

5.2. THE REAL NUMBER SYSTEM

5.2.10 Dedekind Cuts

Definition 5.1. *A Dedekind cut is a pair (A, B) of disjoint nonempty subsets of $\mathbb{Q}$, with $\mathbb{Q} = A \cup B$ and*

$$x < y, \ \forall x \in A, \forall y \in B.$$

Dedekind used these cuts to *define* irrational real numbers. Our irrational numbers are not Dedekind cuts. See Appendix C.

Exercise 5.2.10.1. Suppose that (A, B) is a Dedekind cut. Show that $\sup A \in \mathbb{R}$ and $\sup A = \inf B$.

Exercise 5.2.10.2. Let $x \in \mathbb{R}$. Let $A = \{r \in \mathbb{Q} : r < x\}$ and $B = \{r \in \mathbb{Q} : r \geq x\}$. Show that (A, B) is a Dedekind cut.

Exercise 5.2.10.3. † Show that the function $d : (A, B) \mapsto \sup A$ is a surjection from the set of all Dedekind cuts onto $\mathbb{R}$. Discuss the number of preimages of each $x \in \mathbb{R}$.

5.2.11 Sup of a Sequence

If (a_n) is a sequence of real numbers, then we use the notation

$$\sup_n a_n := \sup \mathrm{im}\, a = \sup\{a_n : n \in \mathbb{N}\},$$

and

$$\inf_n a_n := \inf \mathrm{im}\, a = \inf\{a_n : n \in \mathbb{N}\}.$$

We say that the sequence (a_n) is *bounded above* if $\mathrm{im}\, a$ is bounded above, *bounded below* if $\mathrm{im}\, a$ is bounded below, and *bounded* if $\mathrm{im}\, a$ is bounded.

5.2.12 Sup of a Function

More generally, If $f : E \to \mathbb{R}$ is a real-valued function on some set E, then we use the notation

$$\sup_E f := \sup \mathrm{im}\, f = \sup\{f(x) : x \in E\},$$

and
$$\inf_E f := \inf \mathrm{im} f = \inf\{f(x) : x \in E\}.$$
We say that the function f is *bounded above* if $\mathrm{im} f$ is bounded above, *bounded below* if $\mathrm{im} f$ is bounded below, and *bounded* if $\mathrm{im} f$ is bounded.

5.3 Arithmetical Operations

5.3.1 Unary Minus

Lemma 82. *Let $x \in \mathbb{R}$ and*
$$A = \{-r : r \in \mathbb{Q}, r > x\}.$$
Then A is nonempty and bounded above.

Proof. By the Dense Rationals Axiom, there exist rationals s and t with $s < x < t$. It follows that $-t \in A$, so A is nonempty. Also, if $-r \in A$, then $r > x > s$, so $-r < -s$, and hence $-s$ is an upper bound for A. □

This lemma allows us to extend $-$ from $\mathbb{Q}$ to $\mathbb{R}$ by defining
$$-x = \sup\{-r : r \in \mathbb{Q}, r > x\}.$$

Proposition 83. *Let $x \in \mathbb{R}$. Then x is negative $\Leftrightarrow -x$ is positive.* □

5.3.2 Notation

For $A \subset \mathbb{R}$, we denote
$$-A := \{-x : x \in A\}.$$

Proposition 84. *Let $A \subset \mathbb{R}$ and $b \in \mathbb{R}$. Then b is an upper bound for A if and only if $-b$ is a lower bound for $-A$. Also*
$$\inf A = -\sup(-A).$$
□

5.3. ARITHMETICAL OPERATIONS

Note that the equation in this lemma even holds when $\sup(A)$ is $\pm\infty$.

Corollary 85. *Each nonempty set of real numbers that is bounded below has a greatest lower bound.* □

5.3.3 Absolute Value

We define
$$|x| := \begin{cases} x, & \text{if } x > 0, \\ 0, & \text{if } x = 0, \\ -x, & \text{if } x < 0. \end{cases}$$

Evidently, $|-x| = |x| \geq 0$, for each $x \in \mathbb{R}$.

Exercise 5.3.3.1. Let $A \subset \mathbb{R}$. Then A is bounded if and only if there exists $M > 0$ such that $|x| \leq M$ for each $x \in A$.

5.3.4 Addition

Lemma 86. *Let $x, y \in \mathbb{R}$ and*
$$A = \{r + s : r \in \mathbb{Q}, s \in \mathbb{Q}, r < x, s < y\}.$$

Then:
(1) A is nonempty and bounded above.
(2) Further, if $x, y \in \mathbb{Q}$, then
$$x + y = \sup A.$$

Proof. (1) Take $r, s \in \mathbb{Q}$ with $r < x$ and $s < y$. Then $r + s \in A$, so A is nonempty. Evidently, $x + y$ is an upper bound for A.

(2) Now suppose $x, y \in \mathbb{Q}$. We know that $x + y$ is an upper bound for A. Suppose $b < x + y$. Choose $c \in \mathbb{Q}$ with $b < c < x + y$. Take $n = \text{denom}(x + y - c)$ and $m = \text{num}(x + y - c)$. Then
$$x + y = c = \frac{m}{n} > \frac{2m}{3n}.$$

Let
$$r = x - \frac{m}{3n}, \quad s = y = \frac{m}{3n}.$$

Then $r, s \in \mathbb{Q}$, $r < x$, $s < y$, so $r + s \in A$. But $r + s > c > b$, so b is not an upper bound for A. Thus $x + y$ is the least upper bound of A. □

This lemma allows us to extend $+$ to $\mathbb{R}$ by defining
$$x + y := \sup\{r + s : r \in \mathbb{Q}, s \in \mathbb{Q}, r < x, s < y\},$$
whenever $x, y \in \mathbb{R}$.

Proposition 87. *Let $x, y \in \mathbb{R}$. Then:*

(1)
$$x + y = \sup\{r + s : r \in \mathbb{Q}, s \in \mathbb{Q}, r \leq x, s < y\}.$$

(2)
$$x + y = \sup\{r + s : r \in \mathbb{Q}, s \in \mathbb{Q}, r \leq x, s \leq y\}.$$

(3)
$$x + y = \inf\{r + s : r \in \mathbb{Q}, s \in \mathbb{Q}, r > x, s > y\}.$$

(4) If $x \in \mathbb{Q}$, then
$$x + y = \sup\{x + s : s \in Q, s < y\}.$$

If $x \in \mathbb{Q}$, then
$$x + y = \sup\{x + s : s \in Q, s \leq y\}.$$

□

Proposition 88. *$x > y$ and $z \geq w$ imply $x + z > y + w$.* □

Theorem 89 (Laws of Addition). *Let $x, y, z \in \mathbb{R}$. Then:*
(1) $x + (y + z) = (x + y) + z$.
(2) $x + 0 = x$.
(3) $x + (-x) = 0$.
(4) $x + y = y + x$.

□

5.3. ARITHMETICAL OPERATIONS

5.3.5 Subtraction

Definition 5.2. $x - y := x + (-y), \forall x, y \in \mathbb{R}$

This is consistent with the definition in place for $x, y \in \mathbb{Q}$.

Proposition 90. *Let $x, y \in \mathbb{R}$. Then:*
(1) $x < y \Leftrightarrow y - x$ is positive.
(2) $x > y \Leftrightarrow x - y$ is positive. □

Exercise 5.3.5.1. Suppose that $A \subset \mathbb{R}$ is bounded, and $t \in \mathbb{R}$. Show that there exists $r > 0$ such that $A \subset [t - r, t + r]$.

5.3.6 Triangle Inequality for $\mathbb{R}$

Theorem 91. *Let $a, b \in \mathbb{R}$. Then*

$$|a + b| \leq |a| + |b|.$$

Equality holds only when a and b are not of opposite signs.

Proof. Consider separately the possible cases:
$a \geq 0, b \geq 0$;
$a < 0, b < 0$;
$a \geq 0, b < 0, a + b \geq 0$;
$a \geq 0, b < 0, a + b < 0$;
$a < 0, b \geq 0, a + b \geq 0$;
$a < 0, b \geq 0, a + b < 0$. □

5.3.7 Multiplication

We now extend the multiplication already-defined on the rational numbers to real numbers.

Lemma 92. *Let x and y be positive real numbers. Then*
(1) the set

$$A = \{rs : r \in \mathbb{Q}, s \in \mathbb{Q}, r < x, s < y\}$$

is nonempty and bounded above.
(2) If in addition $x, y \in \mathbb{Q}$, then $xy = \sup A$.

Proof. (1) Exercise.

(2) Evidently, xy is an upper bound for A.

Suppose $b < xy$. Since $xy > 0$, we have $\max(b,0) < xy$, so we may choose $c \in \mathbb{Q}$ with

$$\max(b,0) < c < xy,$$

so that $xy - c > 0$. Since $x + y > 0$, we have $x + y \in \mathbb{Q}^\times$. Take

$$n = \operatorname{denom}\left(\frac{xy-c}{x+y}\right), \quad m = \operatorname{num}\left(\frac{xy-c}{x+y}\right).$$

Then $n, m \in \mathbb{N}$, so

$$\left(\frac{xy-c}{x+y}\right) = \frac{m}{n} \geq \frac{1}{n}.$$

Thus

$$\begin{aligned}\left(x - \frac{1}{n}\right)\left(y - \frac{1}{n}\right) &= xy - \frac{x+y}{n} + \frac{1}{n^2} \\ &> xy - \frac{x+y}{n} > c > b.\end{aligned}$$

Thus b is not an upper bound for A, so xy is $\sup A$. $\square$

This lemma allows us to extend the definition of multiplication from $\mathbb{Q}$ to $\mathbb{R}$, as follows:

Definition 5.3. *If $x \in \mathbb{R}$, we define $x \times 0 := 0 \times x = 0$. If $x > 0$ and $y > 0$, then we define*

$$x \times y := \sup\{r \times s : r \in \mathbb{Q}, s \in \mathbb{Q}, r < x, s < y\}.$$

If $x < 0$ and $y > 0$, we define

$$x \times y := y \times x := -((-x) \times y).$$

If $x < 0$ and $y < 0$, we define $x \times y := (-x) \times (-y)$.

5.3. ARITHMETICAL OPERATIONS

Proposition 93. *Let $x > 0$ and $y > 0$. Then:*

(1)
$$x \times y = \sup\{r \times s : r \in \mathbb{Q}, s \in \mathbb{Q}, r \leq x, s < y\}.$$

(2)
$$x \times y = \sup\{r \times s : r \in \mathbb{Q}, s \in \mathbb{Q}, r \leq x, s \leq y\}.$$

(3)
$$x \times y = \inf\{r \times s : r \in \mathbb{Q}, s \in \mathbb{Q}, r > x, s > y\}.$$

(4) If $x \in \mathbb{Q}$, then
$$xy = \sup\{x \times s : s \in Q, s < y\}.$$

(5) If $x \in \mathbb{Q}$, then
$$xy = \sup\{x \times s : s \in Q, s \leq y\}.$$

Proof. Part (4): Suppose $x \in \mathbb{Q}$. Let
$$A = \{r \times s : r \in \mathbb{Q}, s \in \mathbb{Q}, r < x, s < y\},$$
and and $B = \{x \times s : s \in Q, s < y\}$. Then by definition $xy = \sup A$, and obviously $\sup A \leq \sup B$.

Suppose $xy < \sup B$. Then we may choose $c \in \mathbb{Q}$ with $xy < c < \sup B$, and then choose $b \in B$ with $c < b$, and then choose $s \in \mathbb{Q}$ with $s < y$ and $xs = b$. Then $xy < xs$, so $x \neq 0$, so $y < s$, contradicting the fact that $s < y$. Thus $xy = \sup B$, as claimed.

We leave the other parts as exercises. □

From now on, for all real x and y, we denote $x \times y$ by xy.

Theorem 94. *Let $x, y, z \in \mathbb{R}$. Then*
(1) $x(yz) = (xy)z$,
(2) $xy = yx$.
(3) $x1 = x$.
(4) If $x \neq 0$, then $\exists y \in \mathbb{R}$ such that $xy = 1$.
(5) $x(y + z) = xy + xz$.

Proof. Exercise

Part (3) follows from Part (4) of the last, together with the fact that $x = \sup\{r \in \mathbb{Q} : r < x\}$. □

As before, the reciprocal of a nonzero number is unique, and we denote it x^{-1}.

Proposition 95 (Cancellation Law). *Let $x, y, z \in \mathbb{R}$, $x \neq 0$, and $xy = xz$. Then $y = z$.*

Proof. The proof is the same as for Prop 66. □

The order properties of the arithmetic operations carry over from $\mathbb{Q}$ to $\mathbb{R}$:

Proposition 96. *Let $x, y, z \in \mathbb{R}$. Then:*
(1) $x > y \Rightarrow x - z > y - z$.
(2) If $x > 0$ and $y > 0$, then $x + y > 0$ and $xy > 0$.
(3) If $x > y$ and $z > 0$, then $xz > yz$.
(4) If $x > y$ and $z < 0$, then $xz < yz$.
(5) If $x > y > 0$ or $0 > x > y$, then $y^{-1} > x^{-1}$.
(6) If $x > 0$, then $x^{-1} > 0$.
(7) If $x < 0$, then $x^{-1} < 0$. □

5.3.8 Division

Definition 5.4. *We define the* quotient *of x by y by*
$$\frac{x}{y} := x/y = xy^{-1},$$
whenever $x \in \mathbb{R}$ and $y \in \mathbb{R}^\times$.

Proposition 97. *If $x > 0$ and $y > 0$, then $x/y > 0$.*

Proof. By Part (5) of Proposition 96, $y^{-1} > 0$, so by Part (2) of the same proposition, $x(y^{-1}) > 0$. □

5.4. THE ARCHIMEDEAN PROPERTY

Exercise 5.3.8.1. Let $x > 0$. Show that $1/x = \sup B = \inf C$, where
$$B = \left\{\frac{1}{r} : r \in \mathbb{Q}, r > x\right\},$$
and
$$C = \left\{\frac{1}{r} : r \in \mathbb{Q}, 0 < r < x\right\}.$$

Exercise 5.3.8.2. For real numbers $x \neq y$, show that $\frac{1}{2}(x+y)$ is between x and y (cf. page 72).

5.3.9 Integral Powers

We could now define various integral powers of real numbers, but we prefer to define, in greater generality, these integral powers of complex numbers, and we shall do this in the next chapter. See Subsection 6.1.3.

5.4 The Archimedean Property

5.4.1 The Idea

Archimedes is reported to have said that he could measure any distance, no matter how large, using any measuring stick, no matter how small.

Lemma 98. *Let $a \in \mathbb{Q}$. Then there exists some $n \in \mathbb{N}$ with $n \geq a$.*

Proof. If $a \leq 0$, we may take $n = 1$.
If $a > 0$, then taking $n = \text{num}(a)$, we have $n \in \mathbb{N}$, and
$$n \geq \frac{n}{\text{denom}(a)} = a.$$

$\square$

Theorem 99. *Let $a, b \in \mathbb{R}$ with $a > 0$. Then there exists $n \in \mathbb{N}$ with $na > b$.*

Proof. If $a > b$, we may take $n = 1$. So suppose $b \geq a$. Then $b > 0$, so $b/a > 0$. Choose $r \in \mathbb{Q}$ with $r > b/a$. Then by the lemma we may choose $n \in \mathbb{N}$ with $n \geq r$, and we have $n > b/a$, so $na > b$. □

5.4.2 Floor and Ceiling

The theorem allows us to make the following definition: For $a \in \mathbb{R}$, we let the *ceiling of a* be the integer

$$\lceil a \rceil := \inf\{n \in \mathbb{Z} : a \leq n\}.$$

We also define the *floor of a*:

$$\lfloor a \rfloor := -\lceil (-a) \rceil = \sup\{n \in \mathbb{Z} : a \geq n\}.$$

For instance, $\lfloor 3 \rfloor = 3 = \lfloor \frac{7}{2} \rfloor$ and $\lceil \frac{7}{2} \rceil = 4$.

Exercise 5.4.2.1. † Discuss each relation below in terms of the concepts defined so far. For instance, is it an equivalence relation, and if so, what are the equivalence classes? Is it a partial order? Is it a function, and if so, what are its domain and image, and is it injective? Does it have an inverse function, and if so, what is it?

1. $\{x, y\} \in \mathbb{R} \times \mathbb{R} : x \leq \frac{1}{2} y\}$.

2. $\{(x, y) \in \mathbb{R} \times \mathbb{R} : xy = 1\}$.

3. $\{(x, y) \in \mathbb{R} \times \mathbb{R} : -1 \leq x \leq 1$ and $y > 0$ and $x^2 - 2x + 1 - y = 0\}$.

Exercise 5.4.2.2. † Let $f(x) = x^2 + 1$, $g(x) = x - 2$ for $x \in \mathbb{R}$. Work out

$$f(2 + g(2 + f(2) + g(2))).$$

5.5 Key Properties of the Real Number System

5.5.1 Summary

The set $\mathbb{R} = \{0\} \dot{\cup} \mathbb{R}^{\times}$ of real numbers contains the set $\mathbb{Q}$ of rational numbers. The sum $x + y \in \mathbb{R}$ and product $x \times y \in \mathbb{R}$ (also written $x \cdot y$ or just xy) are (uniquely) defined for each $x, y \in \mathbb{R}$, the negative $-x \in \mathbb{R}$ is defined for each $x \in \mathbb{R}$, and the reciprocal $1/x$ is defined for each nonzero $x \in \mathbb{R}$. There is a positivity concept ('> 0') on $\mathbb{R}$. Restricted to the rational numbers, these operations and this concept agree with those previously discussed in Chapter 4. The following properties hold:

1. **Properties of addition**
    a) Commutative Law: $x + y = y + x \in \mathbb{R}$, $\forall x, y \in \mathbb{R}$.
    b) Associative Law: $x+(y+z) = (x+y)+z$, $\forall x, y, z \in \mathbb{R}$.
    c) $x + 0 = x$, $\forall x \in \mathbb{R}$.
    d) $x + (-x) = 0$, $\forall x \in \mathbb{R}$.

2. **Properties of multiplication**
    a) Commutative Law: $x \times y = y \times x$, $\forall x, y \in \mathbb{R}$.
    b) Associative Law: $x \times (y \times z) = (x \times y) \times z$, $\forall x, y, z \in \mathbb{R}$.
    c) $x \times 1 = x$, $\forall x \in \mathbb{R}$.
    d) If $x \in \mathbb{R}$ and $x \neq 0$, then $x \times (1/x) = 1$.

3. **Distributive Law:**
$$x \times (y + z) = x \times y + x \times z, \quad \forall x, y, z \in \mathbb{R}.$$

4. **Properties of Positivity**
    a) Trichotomy: For each $x \in \mathbb{R}$, exactly one of these is true: $x = 0$; $x > 0$; $-x > 0$.
    b) If $x > 0$ and $y > 0$, then $x + y > 0$, $\forall x, y \in \mathbb{R}$.
    c) If $x > 0$ and $y > 0$, then $x \times y > 0$, $\forall x, y \in \mathbb{R}$.

5. **Completeness:** If $\emptyset \neq E \subset \mathbb{R}$, $b \in \mathbb{R}$, and $x \leq b$, $\forall x \in E$, then there exists $u \in \mathbb{R}$ such that

$$x \leq u \ \forall x \in E,$$

and

$$(x \leq c \ \forall x \in E) \Rightarrow u \leq c.$$

Exercise 5.5.1.1. We have not listed here Axiom 5.3, i.e. the fact that $\mathbb{Q}$ is dense in $\mathbb{R}$. This is because it may be deduced from the properties listed. We used it on the way to *defining* the arithmetic operations on $\mathbb{R}$. But if you just assume you have these operations, $+$ and $\times$, and that they *respect positivity* as in items 4b and 4c above, then you can define $x < y$ to mean $y + (-x) > 0$, and prove Axiom 5.3 by using Completeness.

Verify this, or look it up in some book on basic analysis.

5.6 The Extended Real Numbers

The set of *extended real numbers* is obtained from the set of real numbers by adding two extra atoms, $+\infty$ and $-\infty$.[8] We denote it by

$$\overline{\mathbb{R}} := [-\infty, +\infty] = \mathbb{R} \cup \{-\infty, +\infty\}.$$

We extend the order on $\mathbb{R}$ to a total order on $[-\infty, +\infty]$ by declaring that $-\infty$ is the least, and $+\infty$ the greatest extended real number. Extend unary minus to $[-\infty, +\infty]$ in the obvious way, and we also extend the basic arithmetic binary operations of addition, subtraction, multiplication and division to certain subsets of $[-\infty, +\infty]^2$. For instance, we adopt the conventions that $\frac{1}{\pm\infty} := 0$, but we do not define $\frac{1}{0}$.

---

[8]Strictly speaking, we should state and assume a new axiom, to give us the existence of these atoms. By the way, there is nothing magical or mystical about $\pm\infty$: they are just another two mathematical objects, no more strange than the others, such as 2 and π. They are two among many distinct kinds of infinity that occur in Mathematics, and, in particular, have nothing to do with Theology.

6: The Complex Numbers

By the time they digest the idea of a complex number, many people begin to feel that things are getting out of hand. Complex numbers only became respectable in the nineteenth century, when they were accepted by people of the stature of Gauss and Hamilton. Complex numbers can be modelled as pairs of real numbers, and hence their theory is as consistent as that of the real numbers. Few of those who view them with suspicion nowadays realise that the rot started thousands of years ago, when -1 insinuated itself into the textbooks. People like us know that $\sqrt{-1}$ is no more outrageous than -1: both are pure abstractions. The latter troubles people less only because they became familiar with it in early youth, and have forgotten the strangeness. In many ways, the first mathematician to use -1 was a far greater master than the first to use $\sqrt{-1}$.

6.1 The Complex Number System

6.1.1 The Axioms

Axiom 6.1 (Complex Atoms). *We take as given a particular set $\mathbb{C}$ of atoms, containing $\mathbb{R}$, with a particular element i.*

We denote $\mathbb{C} \sim \{0\}$ by $\mathbb{C}^\times$.

Axiom 6.2 (Complex Arithmetic). *We assume there are binary operations $+$ and $\times$ on $\mathbb{C}$, a unary operation $z \mapsto -z$ on $\mathbb{C}$, and a unary operation $z \mapsto z^{-1}$ on $\mathbb{C}^\times$, extending those already defined on $\mathbb{R}$ and $\mathbb{R}^\times$, and satisfying the following, for each $z, w, t \in \mathbb{C}$:*
(1) $z + (w + t) = (z + w) + t$.
(2) $z + w = w + z$.
(3) $z + 0 = z$.
(4) $z + (-z) = 0$.
(5) $z \times (w \times z) = (z \times w) \times z$.
(6) $z \times w = w \times z$.
(7) $z \times 1 = z$.
(8) *If* $z \neq 0$, *then* $z \times (z^{-1}) = 1$.
(9) $z \times (w + t) = (z \times w) + (z \times t)$.
(10) $i \times i = -1$.
(11) *There exist unique* $x, y \in \mathbb{R}$ *such that* $z = x + (i \times y)$.

The reader will note that $\mathbb{C}$ is just large enough to contain $\mathbb{R}$ and allow us to solve $z^2 = -1$.

6.1.2 Complex Arithmetic

As with the reals, we write $z \cdot w$ or just zw for $z \times w$.

The operations of binary subtraction ($-$), division ($/$) are defined as for the reals:

$$z - w := z + (-w), \quad \forall z, w \in \mathbb{C};$$

$$z/w := z \times (w^{-1}), \quad \forall z \in \mathbb{C}, \forall w \in \mathbb{C}^\times.$$

The arithmetic operations behave in the same way as for the reals, and one has the same cancellation properties.

There are new operations:

6.1. THE COMPLEX NUMBER SYSTEM

Definition 6.1. *If $z = x + iy$, with $x, y \in \mathbb{R}$, then we define the real part of z to be $\Re z := x$, the imaginary part of z to be $\Im z := y$, and the complex conjugate of of z by*

$$\bar{z} := x - iy.$$

Proposition 100. *Let $z, w \in \mathbb{C}$. Then:*
(1) $\Re z = \frac{1}{2}(z + \bar{z})$.
(2) $\Im z = \frac{1}{2i}(z - \bar{z})$.
(3) $\Re(z + w) = \Re z + \Re w$.
(4) $\Im(z + w) = \Im z + \Im w$.
(5) $\bar{z} = z$ if and only if $z \in \mathbb{R}$.
(6) $\bar{z} = -z$ if and only if z is pure imaginary.
(7) $\overline{z + w} = \bar{z} + \bar{w}$.
(8) $\overline{zw} = (\bar{z})(\bar{w})$.
(9) $\overline{z/w} = (\bar{z})/(\bar{w})$, if $w \neq 0$. □

Exercise 6.1.2.1. † Prove $z0 = 0$ for all $z \in \mathbb{C}$.

Exercise 6.1.2.2. † Express the following in the form $a + ib$ for real a and b:
(i) $(2 + 3i)(3 - 2i)$, (ii) $(2 + 3i)/(3 - 2i)$, (iii) $1/i$, (iv) $(1 + i)^3$.

Exercise 6.1.2.3. Show that $z\bar{z}$ is always real, when $z \in \mathbb{C}$.

Exercise 6.1.2.4. † Find all solutions to $\bar{z} = 1/z$.

6.1.3 Integral Exponents

We define integral powers inductively, extending the definition already in place for rationals.

Definition 6.2. *Let $a \in \mathbb{C}$. We define:*

$$\begin{aligned} a^0 &:= 1, \quad \text{if } a \neq 0; \\ a^1 &:= a, \\ a^{n+1} &:= a \times a^n, \quad \forall n \in \mathbb{N}, \end{aligned}$$

and, if $a \neq 0$,

$$a^{-n} := \frac{1}{a^n}, \quad \forall n \in \mathbb{N}.$$

Proposition 101 (Laws of Exponents). *Let $a, b \in \mathbb{C}^\times$, and m and n be integers. Then:*
(1) $(a^m)(a^n) = a^{m+n}$.
(2) $(a^m)^n = a^{mn}$.
(3) $(ab)^n = (a^m)(b^n)$. □

Definition 6.3. *A* polynomial *(over $\mathbb{C}$) is a function $p : \mathbb{C} \to \mathbb{C}$ defined by a formula*

$$p(z) = a_n z^n + \cdots + a_2 z^2 + a_1 z + a_0, \quad \forall z \in \mathbb{C},$$

where $a_0, \ldots, a_n$ are complex constants.
A rational function *is a function r defined by a formula*

$$r(z) = \frac{p(z)}{q(z)}, \quad \forall z \in \mathbb{C} \text{ such that } g(z) \neq 0,$$

where p and q are polynomials.

The a_i are called the *coefficients* of the polynomial. When they are real, p is called a real polynomial, or a polynomial *over* $\mathbb{R}$. When p and q are real polynomials, then p/q is called a rational function *over* $\mathbb{R}$.

We do not define any order on the set of all complex numbers.

Exercise 6.1.3.1. If $x \in \mathbb{R}^\times$, show that $x^2 > 0$.

Exercise 6.1.3.2. Show that the equation $x^2 + 1 = 0$ has no real solution. Hence $i \notin \mathbb{R}$.

6.1.4 Arithmetic on Finite Sequences of Numbers

If $a : \mathbb{N} \to \mathbb{C}$, we define inductively:

(1) $\sum_{j=1}^{1} a_j := a_1,$
(2) $\sum_{j=1}^{n+1} a_j := a_{n+1} + \sum_{j=1}^{n} a_j.$

and

(3) $\prod_{j=1}^{1} a_j := a_1,$
(4) $\prod_{j=1}^{n+1} a_j := a_{n+1} \times \prod_{j=1}^{n} a_j.$

We often write $\sum_{j=1}^{n} a_j$ as $a_1 + \cdots + a_n$, and $\prod_{j=1}^{n} a_j$ as $a_1 \cdots a_n$.

6.1. THE COMPLEX NUMBER SYSTEM

Proposition 102. *Let $a_j, b_j \in \mathbb{C}$ for each $j \in \mathbb{N}$. Then:*

(1)
$$\sum_{j=1}^{n} a_j \pm b_j = \sum_{j=1}^{n} a_j \pm \sum_{j=1}^{n} b_j.$$

(2) If $\alpha \in \mathbb{C}$, then
$$\sum_{j=1}^{n} \alpha b_j = \alpha \sum_{j=1}^{n} b_j.$$

(3)
$$\prod_{j=1}^{n} a_j b_j = \prod_{j=1}^{n} a_j \prod_{j=1}^{n} b_j.$$

(4) If $b_j \neq 0$ for all j, then
$$\prod_{j=1}^{n} \frac{a_j}{b_j} = \frac{\prod_{j=1}^{n} a_j}{\prod_{j=1}^{n} b_j}.$$

Proof. (1)-(3) are proved by induction. (4) is a special case of (3). □

Note that an easy induction yields $\sum_{j=1}^{n} 1 = n$.

6.1.5 Binomial Theorem for Positive Integral Exponents

Theorem 103. *Let $a, b \in \mathbb{C}$ and $n \in \mathbb{N}$. Then*
$$(a+b)^n = \sum_{m=1}^{n} \binom{n}{m} a^m b^{n-m}.$$

Proof. Use induction on n and the law of Pascal's triangle.[1] □

All the following exercises may be tackled using induction.

---
[1] See Section 4.4, page 67.

Exercise 6.1.5.1. Prove
$$1 + 2 + \cdots + n = \frac{n(n+1)}{2}.$$

Exercise 6.1.5.2. Prove
$$1^2 + 2^2 + \cdots + n^2 = \frac{n(n+1)(2n+1)}{6}.$$

Exercise 6.1.5.3. Prove
$$1^3 + 2^3 + \cdots + n^3 = \left(\frac{n(n+1)}{2}\right)^2.$$

Exercise 6.1.5.4. * Find a formula for $1^4 + 2^4 + \cdots + n^4$.

Exercise 6.1.5.5. * Show that for each $m \in \mathbb{N}$,
$$p_n(m) := 1^m + 2^m + \cdots + n^m$$
is a polynomial in n of degree $m + 1$, having leading coefficient $\frac{1}{m+1}$.

Exercise 6.1.5.6. Prove that if $a \in \mathbb{C}$ is not 1, then
$$1 + a + a^2 + \cdots + a^n = \frac{a^{n+1} - 1}{a - 1} = \frac{1 - a^{n+1}}{1 - a}.$$

Exercise 6.1.5.7. Prove
$$1 + 3 + 6 + \cdots + \frac{n(n+1)}{2} = \frac{n(n+1)(n+2)}{6}.$$

Exercise 6.1.5.8. Prove
$$1 + 4 + 10 + \cdots + \frac{n(n+1)(n+2)}{6} = \frac{n(n+1)(n+2)(n+3)}{24}.$$

Exercise 6.1.5.9. * Generalise the last two formulas.

6.2. A PLACE TO STAND

Exercise 6.1.5.10. (Vandermonde[2]) If we adopt the convention that $\binom{n}{k} = 0$ for $n, k \in \mathbb{Z}_+$ with $k > n$, then

$$\binom{m+n}{k} = \sum_{j=0}^{k} \binom{m}{j}\binom{n}{k-j},$$

for $m, n \in \mathbb{Z}_+$ and $k \in \mathbb{Z}$.

Exercise 6.1.5.11. (i) Prove that

$$x^4 + 1 = (x-1)(x+1)(x-i)(x+i),$$

and hence, or otherwise, express $x^4 + 1$ as a product of polynomials over $\mathbb{R}$ of degrees one and two. (ii) Give similar factorisations of $x^4 + 16$.

Exercise 6.1.5.12. Suppose a, b and $c \in \mathbb{C}$ with $|a| = |b| = |c| = 1$ and $a + b + c = 0$. Show that $a\bar{b} + \bar{a}b = -1$, $|a-b|^2 = 3$. (Hence, by symmetry, $|a-b| = |b-c| = |c-a|$.)

6.1.6 The Extended Complex Numbers

The set $\hat{\mathbb{C}}$ of *extended complex numbers* is obtained from the set of complex numbers by adding one extra atom, ∞. This ∞ is not the same as the $+\infty$ of the extended real numbers.[3]

We extend unary minus to $\hat{\mathbb{C}}$ in the obvious way, and we also extend the basic arithmetic binary operations of addition, subtraction, multiplication and division to certain subsets of $\hat{\mathbb{C}}^2$. For instance, we adopt the conventions that $\frac{1}{\infty} := 0$ and $\frac{1}{0} := \infty$.

6.2 A Place to Stand

This is a parting word about what can and cannot be done[4].

---

[2] Alexander-Théophile Vandermonde (1735-96)

[3] As before, strictly speaking, we should state and assume a new axiom, to give us the existence and distinctness of this atom.

[4] Archimedes is supposed to have said: Give me place to stand, and I will move the world (δῶς μοι πᾶ στῶ καὶ τὰν γᾶν κινάσω).

CHAPTER 6. COMPLEX NUMBERS

Once you reached your teens, you probably started to wonder what it was all about, and started to search for some assurances, and thought about what people call The Big Questions. Most people grope towards a satisfactory position of the basic questions of ontology, epistemology, and morality, and seek wisdom, the understanding of the value of things. Some take refuge in the 'certainties' of the natural sciences. Some realise that these are not all that certain, and take refuge in the 'absolute certainties' of mathematics. If you take these seriously, then you have to insist on complete rigour in mathematical argument. You probably read this book because you were not completely happy with the starting-point of your first analysis course, and want to start 'further down'.

This urge to get to the bottom of things, to have a firm foundation, is deep in us. The Good Book warns about the perils of building on sand. As small children, we relate to the little pig who built his house with solid bricks and withstood the Big Bad Wolf. Unfortunately, even mathematics is a bit like Terry Pratchett's Diskworld. The Diskworld stands upon the shoulders of four huge elephants, who stand upon the great turtle A'Tuin, but unfortunately A'Tuin stands on nothing, or more accurately swims in nothing. In mathematics, we cannot prove the absolute consistency of the axioms of a complete ordered field. This means that someone might someday prove, using these axioms, that $0 = 1$ so that *everything is true and everything is false*. Of course, we hope it ain't so. The story we tell in this book provides this assurance: if real analysis is self-contradictory, then a much more basic theory is also self-contradictory.

This is some comfort, but you cannot be sure. You might be inclined to say: I'm sure that my intuitions about counting are sound, so I really *believe* that Peano's arithmetic is consistent. But it's not so clear. There is some evidence that our beliefs about counting are not innate. I'm told that small children do not assume that the number of wooden blocks in a box remains invariant while the box is closed. They eventually come to believe it, by some process. Few of the things we count are as durable as wooden blocks, hardly any things are really durable,

6.2. A PLACE TO STAND

and there are practical limits to our counting abilities. So to assert the consistency of Peano's postulates on the basis of real experience is to fall into the same fallacy as the physicist who thinks that you can prove Green's Theorem by appealing to experimental observation of fluids. The argument is essentially circular in both cases. It takes the form: I assume my theory describes reality, reality is really there and behaves as my theory says, hence my theory is consistent. This argument is worthless.

A: Propositional Calculus

A.1 Logical Operations

A.1.1 Propositions

A proposition is a thing that has precisely one of two *truth values:* true, or false. We say it is true or false, according to the value.

Given a proposition, p, one gets a new proposition, denoted $\neg p$, called *the negation of p*, and pronounced *not-p*, which is true if p is false, and false if p is true. One thinks of $\neg$ as an operator that reverses the truth value of any given proposition to which it is applied. Concretely, one may think of $\neg p$ as obtained from p by prefixing the words "It is false that" to the statement of p.

p	q	$\neg p$	$p \wedge q$	$p \vee q$
true	true	false	true	true
true	false	false	false	true
false	true	true	false	true
false	false	true	false	false

Table A.1: Logical Operators

Given two propositions, p and q, one may form a new proposition denoted $p \wedge q$, called *the conjunction of p and q*, and pronounced *p and q*, which is true only when both p and q are true, and is otherwise false. One may also form another new proposition $p \vee q$, called *the disjunction of p and q*, and pronounced *p or q*, which is true only when at least one of p and q is true, and is false only when both are false. The manner in which the logical operators $\neg$, $\wedge$ and $\vee$ work is summarised in Table A.1.

A.2 Logical Rules

A.2.1 Compound Propositions

A compound proposition is one that is composed from simpler propositions by using some combination of the operators $\neg$, $\wedge$ and $\vee$. The method for manipulating compound propositions is called the elementary propositional calculus. Here are the basic rules:

Definition A.1. *Two propositions p and q are called* equivalent *(denoted $p \Leftrightarrow q$) if they have the same truth value.*

A.2.2 Notation

It is commonly done to use the letter T, or the number 1 to denote some fixed true proposition, and F, or 0 to denote a particular false one. (It doesn't matter, in practice, which particular propositions are used. For instance, one may take T to be the statement $1 = 1$, and F to be the statement $1 \neq 1$.) We use T and F below.

Proposition 104. *Let p, q, and r be propositions. Then*
(1) $\wedge$ is commutative: $p \wedge q \Leftrightarrow q \wedge p$.
(2) $\wedge$ is associative: $p \wedge (q \wedge r) \Leftrightarrow (p \wedge q) \wedge r$.
(3) $p \wedge T \Leftrightarrow p$.
(4) $p \wedge F \Leftrightarrow F$.
(5) $\vee$ is commutative: $p \vee q \Leftrightarrow q \vee p$.
(6) $\vee$ is associative: $p \vee (q \vee r) \Leftrightarrow (p \vee q) \vee r$.
(7) $p \vee T \Leftrightarrow T$.
(8) $p \vee F \Leftrightarrow p$.
(9) $\wedge$ distributes over $\vee$: $p \wedge (q \vee r) \Leftrightarrow (p \wedge q) \vee (p \wedge r)$.
(10) $\vee$ distributes over $\wedge$: $p \vee (q \wedge r) \Leftrightarrow (p \vee q) \wedge (p \vee r)$.
(11) The Law of the Excluded Middle: $p \vee \neg p \Leftrightarrow T$.
(12) The Law of Contradiction: $p \wedge \neg p \Leftrightarrow F$.
(13) One De Morgan Law: $\neg(p \wedge q) \Leftrightarrow (\neg p) \vee (\neg q)$.
(14) The other De Morgan Law: $\neg(p \vee q) \Leftrightarrow (\neg p) \wedge (\neg q)$.

Outline Proof. The truth value of a compound proposition depends only on the truth values of the component propositions, and not on their detailed statements. Thus each of these rules may be verified by checking all possible combinations of truth values for whichever of the variables p, q and r occur in the rule. This involves at most 8 cases for each rule, and is a routine exercise. The best way to do it is to make a table for each rule. For instance, in the case of 13, one has the table:

p	q	$p \wedge q$	$\neg(p \wedge q)$	$\neg p$	$\neg q$	$(\neg p) \vee (\neg q)$
T	T	T	F	F	F	F
T	F	F	T	F	T	T
F	T	F	T	T	F	T
F	F	F	T	T	T	T

Table 61: First De Morgan Law

The fact that the fourth and seventh columns are identical means that in all cases $\neg(p \wedge q)$ and $(\neg p) \vee (\neg q)$ have the same truth value, i.e. are equivalent, as asserted. □

In the text, we almost always use the ordinary English words *and* for $\wedge$ and *or* for $\vee$. For $\neg$, we express statements such as $\neg(x = 3)$ by writing $x \neq 3$ or "$x = 3$ fails", etc. Note in particular that we *always* use the word *or* to mean the so-called *inclusive or*. Thus "p or q" is always equivalent to $p \vee q$, and not (always) to the *exclusive or*: $(p \vee q) \wedge \neg(p \wedge q)$. The latter is often denoted p XOR q.

It may be helpful to note that if one uses 1 and 0 to denote T and F, then the algebra of $\wedge$ is precisely the same as ordinary multiplication in $\mathbb{Z}$. The algebra of $\vee$ corresponds to what is called *Boolean addition*, which is different from ordinary addition, even modulo 2. Ordinary addition modulo 2 corresponds to XOR, so we have the formula

$$p \vee q \Leftrightarrow (p + q) + (p \times q) \mod 2.$$

B: Complete List of Axioms

We gather here, for ease of reference, the axioms assumed at various points in the text. We have added a version of the Axiom of Choice, although it is not needed for the first analysis course.

B.1 Sets

[Membership] 1.1:
If A is a class, then each member of A is a class or an atom.

[Extent, or Class Equality] 1.2:
If A and B are classes, then

$$(A = B) \Leftrightarrow (a \in A \Leftrightarrow a \in B).$$

[Classification] 1.3:
For each predicate $\phi(x)$ there exists a class C such that

$$(a \in C) \Leftrightarrow (a \text{ is a set or an atom}) \text{ and } \phi(a)).$$

[Union] 1.4:
If A and B are sets, then so is $A \cup B$.

[Power Set] 1.5:
If A is a set, then there is a set S such that

$$x \subset A \Leftrightarrow x \in S.$$

[Amalgamation] 1.6:
If $\mathfrak{A}$ is a set, then so is $\bigcup \mathfrak{A}$.

[Ordered Pairs] 1.7:
For each $a \in \mathcal{U}$ and $b \in \mathcal{U}$, there exists an atom (a, b), such that

$$(a, b) = (c, d) \Leftrightarrow (a = c \wedge b = d).$$

whenever c and d are sets or atoms.

[Substitution] 1.8:
If f is a function on a set A, then im f is also a set.

[Choice]
If A is a set and f is a function on A such that $f(x)$ is a nonempty set for each $x \in A$, then there is a function g on A such that $g(x) \in f(x)$ for each $x \in A$.

B.2 Natural Numbers

[Natural Numbers] 2.1:
There is a particular set $\mathbb{N} \subset \mathcal{A}$ of atoms, and a particular element $1 \in \mathbb{N}$.

[Successor] 2.2:
There is a particular function $s : \mathbb{N} \to \mathbb{N}$ that is a bijection from $\mathbb{N}$ onto $\mathbb{N} \sim \{1\}$.

[Induction] 2.3:

$$\left\{ \begin{array}{ccc} A & \subset & \mathbb{N} \\ 1 & \in & A \\ n \in A & \Rightarrow & s(n) \in A \end{array} \right\} \Rightarrow (A = \mathbb{N}).$$

B.3 Integers

[Integers] 3.1:
We take as given a set $\mathbb{Z}$ of atoms, called integers, containing $\mathbb{N}$. We assume that there is a distinguished atom $0 \in \mathbb{Z}$, and that there is an injective function $- : \mathbb{N} \to \mathbb{Z}$.

[Trichotomy] 3.2:
If $m \in \mathbb{Z}$, then exactly one of these is true: (1) $m = 0$, (2) $m \in \mathbb{N}$, or (3) $\exists! n \in \mathbb{N}$ with $m = -n$.

B.4 Rationals

[Rationals] 4.1
We take as given a set $\mathbb{Q}$ of atoms, called rational numbers, containing the set $\mathbb{Z}$ of integers, and equipped with a binary operation $\times$, extending the operation $\times$ already defined on $\mathbb{Z}$, and such that
(1) (Associative) $(x \times y) \times z = x \times (y \times z)$, for all $x, y, z \in \mathbb{Q}$.
(2) (Commutative) $x \times y = y \times x$, for all $x, y \in \mathbb{Q}$.
(3) (Identity) $x \times 1 = x$, for all $x \in \mathbb{Q}$.
(4) (Inverse) For each nonzero $x \in \mathbb{Q}$, there exists a $y \in \mathbb{Q}$ such that $x \times y = 1$.
(5) (Fraction) For each $x \in \mathbb{Q}$, there exists $y \in \mathbb{N}$ such that $x \times y \in \mathbb{Z}$.

B.5 Reals

[Real Atoms] 5.1:
We take as given a set $\mathbb{R}$ of atoms, called real numbers, containing the set $\mathbb{Q}$ of rational numbers.

[Order] 5.2:
We assume that there is a strict order $>$ on $\mathbb{R}$, extending the strict order $>$ on $\mathbb{Q}$.

[Dense Rationals] 5.3:
If $a, b \in \mathbb{R}$ and $a < b$, then there exist $q, r, s \in \mathbb{Q}$ with $q < a <$

$r < b < s$.

[Completeness] 5.4:
Each nonempty set of real numbers that is bounded above has a least upper bound.

B.6 Complex Numbers

[Complex Atoms] 6.1:
We take as given a particular set $\mathbb{C}$ of atoms, containing $\mathbb{R}$, with a particular element i.

[Complex Arithmetic] 6.2:
We assume there are binary operations $+$ and $\times$ on $\mathbb{C}$, a unary operation $z \mapsto -z$ on $\mathbb{C}$, and a unary operation $z \mapsto z^{-1}$ on $\mathbb{C}^\times$, extending those already defined on $\mathbb{R}$ and $\mathbb{R}^\times$, and satisfying the following, for each $z, w, t \in \mathbb{C}$:
(1) $z + (w + t) = (z + w) + t$.
(2) $z + w = w + z$.
(3) $z + 0 = z$.
(4) $z + (-z) = 0$.
(5) $z \times (w \times z) = (z \times w) \times z$.
(6) $z \times w = w \times z$.
(7) $z \times 1 = z$.
(8) If $z \neq 0$, then $z \times (z^{-1}) = 1$.
(9) $z \times (w + t) = (z \times w) + (z \times t)$.
(10) $i \times i = -1$.
(11) There exist unique $x, y \in \mathbb{R}$ such that $z = x + iy$

C: Relative Consistency

In this appendix we show how to see that the axioms we have introduced for the real and complex numbers are consistent, *provided we assume that the axioms for sets and natural numbers are consistent.*

This kind of thing is called *relative consistency*. We begin with some general principles.

C.1 Mathematical Theories

C.1.1 The Ingredients

A mathematical theory has five ingredients:

1. Undefined terms

2. Definitions of terms

3. Axioms

4. Rules about logical deduction

5. Theorems.

The definitions are not essential ingredients; they are just a tool for abbreviating the statement of theorems. The theorems of the theory are the statements that may be proved about the undefined terms by using the axioms and the logical rules.

C.1.2 Consistency

A given theory is *consistent* if there is no theorem t such that its negation $\neg t$ is also a theorem.

Suppose T_1 and T_2 are two theories. A *model* of T_2 inside T_1 is a class M of terms (defined or undefined) of T_1 and a bijection I from the class of undefined terms of T_2 onto M (called an *interpretation of T_2 inside T_1*) such that each axiom of T_2 becomes a theorem of T_1 when each term t of T_2 is replaced by its image $I(t)$ in T_1. In other words, all the terms of T_2 can be interpreted as terms of T_1, and then all the axioms (and hence all the theorems) of T_2 become provable statements of theory T_1.

If we assume that theory T_1 is consistent, then we can prove that another theory T_2 is consistent by showing that there is a model of T_2 inside T_1. For if there were a contradiction in T_2, there would be a proof of a contradiction in T_1 as well.

C.2 Sets and Relations

C.2.1 Atoms

We treat 2 as an 'atom'. Set theory with many atoms may be modelled in set theory with only one (— namely the empty set), by the simple device of selecting a class N of sets to designate as new 'atoms', and cutting away their members in the membership graph, by defining a new 'belongs' by

$$a \in_1 b \Leftrightarrow \left\{ \begin{array}{cc} a & \in \ b, \\ b & \notin \ N. \end{array} \right\}$$

Thus we risk no loss of consistency by allowing atoms, and we gain the possibility of adhering closer to the original version of various mathematical concepts. The task of laying a foundation for analysis is not well done if the concepts are so altered that Weierstrass could not recognise his own bit.

C.2.2 Ordered Pairs

It has been traditional to *define*

$$(a, b) := \{\{a\}, \{a, b\}\}$$

C.2. SETS AND RELATIONS

when a and b are sets of sets. This is motivated by the fact that if a and b are sets and

$$P = \{\{a\}, \{a, b\}\},$$

then we have

$$\begin{aligned} a &= \bigcap\bigcap P \\ b &= (\bigcap\bigcup P) \cup ((\bigcup\bigcup P) \sim (\bigcup\bigcap P)) \end{aligned}$$

as you may readily verify, so that P uniquely determines the sets a and b, in order.

We could modify this to deal with sets or atoms by using

$$P = \{\{\{a\}\}, \{\{a\}, \{b\}\}\}$$

noting that

$$\{a\} = \{b\} \Leftrightarrow a = b \; \forall a, b \in \mathcal{U}.$$

But cunning and striking as this may be, it is only one of many possible ways to invent a class formula from which a and b may be uniquely recovered, and it has little to recommend it unless we insist that everything in sight be a class, or unless we are engaged in establishing the relative consistency of a set theory involving ordered pairs.

Our procedure is to do the obvious thing, and define (a, b) as 'bracket-a-comma-b-bracket', a new thing associated to a and b, not a class, but an 'atom-with-structure'

This procedure is as consistent as the other. Showing this is not our job in this book. It is work for logicians, not analysts. See, for example, the work of P. Aczel for details: http://www.cs.manchester.ac.uk/~petera/. Incidentally, it is not clear whether one could consistently assume there is an atom (a, b) when a and b are any classes, so as to ensure

$$(a, b) = (c, d) \Leftrightarrow (a = c \wedge b = d).$$

On the face of it, there appears no reason why not, but one learns to be cautious about verbal formalism.

C.3 Integers

This theory can be modelled without the new atoms (and hence is as consistent as the theory of natural numbers), by interpreting integers as equivalence classes of pairs

$$(m, n) \in \mathbb{N} \times \mathbb{N}$$

under the equivalence relation defined by saying that $(m, n) \approx (m', n')$ if and only if there exist $p, q \in \mathbb{N}$ such that $m+p = m'+q$ and $n+p = n'+q$. This relation is the equivalence relation whose classes are the level sets of the function

$$\begin{cases} \mathbb{N} \times \mathbb{N} & \to \quad \mathbb{Z} \\ (m, n) & \mapsto \quad m - n. \end{cases}$$

C.4 Rationals

This theory of rationals can be modelled without new atoms (and hence is as consistent as the theory of integers), by interpreting rationals as equivalence classes of pairs of integers $(m, n) \in \mathbb{Z} \times \mathbb{N}$, under the equivalence relation $\approx$ defined by setting $(m, n) \approx (m', n')$ whenever $m \times n' = m' \times n$.

C.5 Reals

It is possible to *construct* a model of the theory of real numbers within the theory of the rational numbers. This was first done by Dedekind's 'cut' construction (described in his little book [1], or in his article reprinted in Newman's World of Mathematics [7], see also Subsection 5.2.10. Another method involves equivalence classes of Cauchy sequences of rationals.

C.6 Complex Numbers

Complex numbers can be modelled as pairs of real numbers, and hence their theory is as consistent as that of the real numbers.

D: Redundancy in the Axioms

In this appendix we discuss the extent to which some of the axioms, or more accurately some parts of the axioms are redundant, in the sense that if removed they can still be proven as propositions.

In order to prove that a given axiom is not redundant, it suffices to give an interpretation, or model, in which the given axiom is false and all the other axioms are true.

We discuss the axioms of each chapter subject to the assumption that all the axioms of each previous chapter hold. In constructing models for a subset of the axioms, it is convenient to have to hand models that satisfy all the axioms. We denote such models of $\mathbb{N}$, $\mathbb{Z}$, $\mathbb{Q}$, $\mathbb{R}$ and $\mathbb{C}$ and their associated operations $+$, $-$, $\times$ and $/$ and special numbers 0, 1 and i by $\hat{\mathbb{N}}$, $\hat{\mathbb{Z}}$, $\hat{\mathbb{Q}}$, $\hat{\mathbb{R}}$ and $\hat{\mathbb{C}}$, $\hat{+}$, $\hat{-}$, $\hat{\times}$ and $\hat{/}$, and $\hat{0}$, $\hat{1}$ and $\hat{i}$. For instance, this means that the interpretation

$$\mathbb{N} = \hat{\mathbb{N}}, + = \hat{+}, 1 = \hat{1}, s : n \mapsto n\hat{+}1$$

makes the axioms 2.1-2.3 true, and hence all the propositions of Chapter 2 true, assuming the axioms 1.1-1.8 of Chapter 1.

D.1 Sets

D.1.1 Axiom of Infinity

We have not included an "axiom of infinity" in Chapter 1. One form of this axiom states: There exists a set A such that $\emptyset \in A$ and

$$B \in A \Rightarrow B \cup \{B\} \in A.$$

This axiom is often used to construct a model of the natural numbers $\mathbb{N}$ and the successor function s within set theory. In our treatment, the axiom of infinity becomes a potential theorem by Chapter 2, since it may be proven using our axioms about sets and natural numbers. This is why we omitted it from our assumptions.

Exercise D.1.1.1. Prove the Axiom of infinity, using Axioms 1.1-1.8 and 2.1-2.3.

D.2 Natural Numbers

D.2.1

Axiom 2.1 will fail if we interpret $\mathbb{N}$ as some proper class, or as a set whose elements are not all atoms.

D.2.2

If we do not assume Axiom 2.2, there may not be any infinite sets. We can make a model of set theory satisfying all the axioms of Chapter 1, by interpreting 'set' as 'finite set'. In such a model there is no way to choose 1 and $\mathbb{N}$ so that Axioms 2.1 and 2.2 hold.

D.2.3

We can make a model in which Axioms 2.1 and 2.2 hold, but Axiom 2.3 (Induction) fails by defining $1 = \hat{1}$, $\mathbb{N} = \hat{\mathbb{N}} \cup (\{1\} \times \hat{\mathbb{Z}})$, and defining the successor by $s(n) := n \hat{+} 1$ for all $n \in \hat{\mathbb{N}}$ and $s((1, n)) := (1, n \hat{+} 1)$ for all $n \in \hat{\mathbb{Z}}$. This creates a two-sided infinite chain of 'natural numbers' that cannot be reached from 1 by repeatedly taking successors.

D.3 Integers

D.3.1

If we interpret $\mathbb{Z} = \hat{\mathbb{N}}$, then Axiom 3.1 fails, but Axiom 3.2 holds. So we cannot dispense with 3.1.

D.3.2

Given Axiom 3.1, Axiom 3.2 may fail in various ways. It is possible to find models in which for some m none of the three conditions (1), (2), or (3) hold, and it is also possible to find models in which for some m more than one of these three conditions holds:

(1) Intepret $\mathbb{N}$ as $\hat{\mathbb{N}}$, $\mathbb{Z}$ as $\hat{\mathbb{N}}$, 0 as $\hat{1}$, and $-n$ as n for each $n \in \mathbb{N}$. Then Axiom 3.1 holds, but Axiom 3.2 fails. Indeed, for each $m \in \mathbb{Z}$ at least two of the conditions (1), (2), (3) holds.

(2) Pick some atom a (such as $\frac{1}{2} \in \hat{\mathbb{Q}}$) that does not belong to $\hat{\mathbb{Z}}$, and interpret $\mathbb{Z}$ as $\hat{\mathbb{Z}} \cup \{a\}$. Take $0 = \hat{0}$. Then Axiom 3.1 holds, but Axiom 3.2 fails, because the element a satisfies none of (1), (2), or (3).

D.4 Rationals

D.4.1 Axiom 4.1 item 1

Interpret $\mathbb{Q}$ as $\hat{\mathbb{Z}} \cup \{\frac{1}{n} : n \in \hat{\mathbb{Z}}, n \neq 0\}$ (where $1/\pm 1$ is identified with ± 1). Define $\times$ by setting $m \times n := m \hat{\times} n$ for $m, n \in \hat{\mathbb{Z}}$,

$$\frac{1}{m} \times \frac{1}{n} := \frac{1}{m \times n}, \ \forall \text{ nonzero } m, n \in \hat{\mathbb{Z}},$$

and

$$m \times \frac{1}{n} := \frac{1}{n} \times m := \text{sign}(m \times n)$$

whenever $m \in \hat{\mathbb{Z}}$ and $0 \neq n \in \hat{\mathbb{Z}}$, where

$$\text{sign}(n) = \begin{cases} +1 & , & m > 0, \\ -1 & , & m < 0, \\ 0 & , & m = 0. \end{cases}$$

Exercise D.4.1.1. † Show that properties 2, 3, 4, 5 of Axiom 4.1 hold, but property 1 fails.

D.4.2 Axiom 4.1 item 2

This item, the commutative law of multiplication is redundant.

Proposition 105. *Assume the axioms 1.1-1.8, 2.1-2.3, 3.1, 3.2, and items 1, 3, 4 and 5 of Axiom 4.1. Then $x \times y = y \times x$ for all $x, y \in \mathbb{Q}$.*

We break the proof into steps.
Step 1. The cancellation law works.

Exercise D.4.2.1. Check that Proposition 66 holds, i.e. that its proof does not depend on the commutativity of multiplication.

Step 2. $n \times x = x \times n$ for all $n \in \mathbb{N}$ and $x \in \mathbb{Q}$.

Proof. Fix $n \in \mathbb{N}$ and $x \in \mathbb{Q}$. By item 5, we may pick $m \in \mathbb{N}$ with $x \times m \in \mathbb{Z}$. Then $(n \times x) \times m = n \times (x \times m) = (x \times m) \times n = x \times (m \times n) = x \times (n \times m) = (x \times n) \times m$.

Exercise D.4.2.2. Justify each equation.

By cancellation, $n \times x = x \times n$, as claimed.

Step 3. Fix $x, y \in \mathbb{Q}$. By item 5 we may choose $m, n \in \mathbb{N}$ with $x \times m \in \mathbb{Z}$ and $y \times n \in \mathbb{Z}$.

Exercise D.4.2.3. Justify the following:
$(x \times m) \times (y \times n) = x \times (m \times (y \times n)) = x \times ((m \times y) \times n) = x \times ((y \times m) \times n) = (x \times y) \times (m \times n)$.

Similarly, $(y \times n) \times (x \times m) = (y \times x) \times (n \times m)$, and this equals $(y \times x) \times (m \times n)$.

But $(x \times m) \times (y \times n) = (y \times n) \times (x \times m)$, so $(x \times y) \times (m \times n) = (y \times x) \times (m \times n)$, and cancellation yields $x \times y = y \times x$.

This proves the proposition.

D.4. RATIONALS

D.4.3 Axiom 4.1 item 3

Take some element $a \in \hat{\mathbb{R}}$ (or any other set) that does not belong to $\mathbb{Q}$, and interpret $\hat{\mathbb{Q}}$ as $\hat{\mathbb{Q}} \cup \{a\}$. Extend $\hat{\times}$ on $\hat{\mathbb{Q}}$ to a multiplication $\times$ on $\mathbb{Q}$ by defining $q \times a := a \times q := q$ for all $q \in \hat{\mathbb{Q}}$ and $a \times a := 1$. You could picture this as treating a as a distinct version of 1 for the purposes of multiplication, with the proviso that any multiplication lands you in $\hat{\mathbb{Q}}$.

Exercise D.4.3.1. Check that all parts of Axiom 4.1 hold apart from item 3.

Item 3 fails, because $1 \times a = 1 \neq a$.

Item 3 could be replaced by a weaker axiom:

Exercise D.4.3.2. † Assume all parts of Axiom 4.1 except item 3, and assume that natural numbers can be cancelled, i.e. if $x, y \in \mathbb{Q}$, $n \in \mathbb{N}$ and $x \times n = y \times n$, then $x = y$. Show that $x \times 1 = x$ for all $x \in \mathbb{Q}$.

D.4.4 Axiom 4.1 item 4

Item 4 cannot be deduced from the rest. If we interpret $\mathbb{Q}$ as $\hat{\mathbb{Z}}$, and $+$ and $\times$ as $\hat{+}$ and $\hat{\times}$ on $\hat{\mathbb{Z}}$, then items 1, 2, 3 and 5 hold, but not item 4.

Exercise D.4.4.1. Check this.

D.4.5 Axiom 4.1 item 5

Item 5 cannot be deduced from the rest. If we interpret $\mathbb{Q}$ as $\hat{\mathbb{Q}}(\hat{i})$, namely the set of complex numbers having ordinary rational real and imaginary parts, and $+$ and $\times$ as the complex $\hat{+}$ and $\hat{\times}$, then items 1, 2, 3 and 4 hold, but not item 5.

Exercise D.4.5.1. Check this.

D.5 Reals

We confine ourselves to noting how each additional axiom adds new information not deducible from its predecessors.

The first two axioms about the real numbers, Axioms 5.1 and 5.2 are clearly independent of previous axioms. For the first, take any model of $\mathbb{R}$ that has elements that are not atoms. For the second, just do not assume any order on $\mathbb{R}$.

D.5.1 Axiom 5.3

The nonstandard real numbers introduced by A. Robinson in the 1960's provide an ordered model of $\mathbb{R}$ that is non-archimedean.

D.5.2 Axiom 5.4

The Axiom of Completeness fails if we interpret $\mathbb{R}$ as $\hat{\mathbb{Q}}$.

D.6 Complex Numbers

There is considerable redundancy in the properties listed in Axiom 6.2. The eleven properties may be reduced to no more than seven, namely (1),(2),(5),(6),(9),(10), and (11).

We shall show that (2) (commutativity of $+$) may be relaxed a little, and that the pair (6)∧(9) may be replaced by a pair (6w)∧(9s), consisting of a weaker form of (6) (commutativity of $\times$) and a stronger form of (9) (distributivity of $\times$ over $+$).

These variant properties are:

(2w) $x + i \times y = i \times y + x$, $\forall x, y \in \mathbb{R}$.
(6w) $x \times i = i \times x$, $\forall x \in \mathbb{R}$.
(9s) $z \times (w+t) = (z \times w) + (z \times t)$ and $(w+t) \times z = (w \times z) + (t \times z)$ for all $z, w, t \in \mathbb{C}$.

Obviously, (2) implies (2w), (6) implies (6w) and (9s) implies (9).

We shall use the BODMAS rule, giving multiplication precedence over addition, to reduce the number of brackets.

D.6. COMPLEX NUMBERS

D.6.1 Property 6.2(2w)

Property 6.2(2), the commutativity of addition, may be relaxed somewhat:

Proposition 106. *Assume properties (1), (2w), (9), (11). Then property (2) holds.*

Proof. Let $z, w \in \mathbb{C}$. By property (9) we may choose $x, y, u, v \in \mathbb{R}$ with $z = x + i \times y$ and $w = u + i \times v$. Then we have $z+w = (x+i\times y)+(u+i\times v) z+w = (x+i\times y)+(i\times v+u) = x+((i\times y+i\times v)+u) = x+(i\times(y+v)+u) = x+(u+i\times(y+v)) = (x+u)+i\times(y+v)$. Similarly, we have $w + z = (u+x)+i \times (v+y)$. Then, using the commutativity of $+$ on $\mathbb{R}$, we obtain $z + w = w + z$, as required.

D.6.2 Zero

We now show (among other things) that part (3) of Axiom 6.2 is redundant.

Proposition 107. *Suppose (all the other axioms and) parts (1), (2), (5), (9), (10) and (11) of Axiom 6.2. Then*

1. $(-1) \times i = i \times (-1)$;
2. $1 \times i = i \times 1$;
3. $0 = 0 + i \times 0$;
4. $z + 0 = z$ for all $z \in \mathbb{C}$.

Proof. 1. $(-1) \times i = (i \times i) \times i = i \times (i \times i) = i \times (-1)$.
2. Similar, using $1 = i \times i \times i \times i \times i$.
3. By Axiom 6.2(9), there are unique $x, y \in \mathbb{R}$ such that $0 = x + iy$. We then have $0 = (-1) \times 0 = (-1) \times (x + i \times y) = (-1) \times x + (-1) \times (i \times y) = (-x) + ((-1) \times i) \times y = (-x) + (i \times (-1)) \times y = (-x) + i \times ((-1) \times y) = (-x) + i \times (-y)$. Since the representation of 0 as $x + i \times y$ is unique, we conclude that $x = -x$ and $y = -y$, hence $x = 0 = y$. Thus $0 = 0 + i0$.
4. We may write $z = x+iy$, and then by part 3, $z+0 = (x+iy) + (0 + i0)$, and using the associativity 6.2(1) and commutativity 6.2(2) of $+$, and distributivity 6.2(9) this equals $(x + 0) + i \times (y + 0) = x + i \times y = z$.

D.6.3 Minus

The assumption in Axiom 6.2 of the existence of the operation $z \mapsto -z$ and its associated property 6.2(4) are redundant. If we assume the rest, then we may define the operation by setting

$$-(x + i \times y) := (-x) + i \times (-y), \ \forall x, y \in \mathbb{R},$$

and verify property (4).

Exercise D.6.3.1. Check this, stating which properties you are using at each step. Note, specifically, that Axiom 6.4(8) need not be used.

D.6.4 Reciprocal

Similarly, the existence of the operation $z \mapsto z^{-1}$ on $\mathbb{C}^\times$ and its associated property (8) are also redundant. Note that if $x + i \times y \in \mathbb{C}^\times$, then not both x and y can be zero.

Exercise D.6.4.1. Assuming (all other axioms) and all parts of Axiom 6.2 except items (4) and (8), prove that if we define

$$(x + i \times y)^{-1} := (x^2 + y^2)^{-1} \times (x - i \times y)$$

then (8) holds.

Thus both minus and reciprocal and the properties (4) and (8) may be removed from Axiom 6.2 without losing anything essential.

D.6.5 Property 6.2(6w)

Property (6) of Axiom 6.2 may be relaxed.

Proposition 108. *Assume properties (1), (2), (5), (6w), (9s), (10) and (11) of Axiom 6.2 and (6w). Then property (6) holds. Conversely, if we assume properties (6) and (9), then property (9s) holds.*

D.6. COMPLEX NUMBERS

Proof. Fix $z, w \in \mathbb{C}$. By (11), we may choose $x, y, u, v \in \mathbb{R}$ such that $z = x + i \times y$ and $w = u + i \times v$. Then by (1), (2), (5), (6w), (9s), (10) and (11) and the commutativity of $\times$ on $\mathbb{R}$, we have $z \times w = (x+i\times y) \times (u+i\times v) = x\times(u+i\times v)+(i\times y)\times (u+i\times v) = xu+x\times(i\times v)+(i\times y)\times u+(i\times y)\times(i\times v) = xu+(x\times i)\times v+i\times(yu)+i\times((y\times i)\times v) = xu+(i\times x)\times v+i\times(yu)+i\times(i\times y)\times v) = xu+i\times(xv)+i\times(yu)+i\times(i\times y)\times v) = xu+i\times(xv)+i\times(yu)+(-yv) = ux+i\times(vx)+i\times(uy)+(-vy) = ux+i\times(uy)+i\times(vx)+(-vy) = (u+i\times v)\times(x+i\times y) = w\times z$. This proves the first statement. The second is obvious.

D.6.6 Property 6.2(7)

Property (7) is redundant.

Exercise D.6.6.1. † Assume Properties (5), (9s), (11), of Axiom 6.2. Prove that $z \times 1 = z$ for all $z \in \mathbb{C}$.

D.6.7 Property 6.2(11)

It is sufficient to assume the existence part of Property (11), given the others. We have noted (cf. exercises in Subsection 6.1.3) that i is not a real number, and this can be proven just by using the axioms of the Real number system and Property 6.2(10). Assuming, in addition, the associativity and commutativity of complex multiplication, one can show by squaring it that no nonzero pure imaginary number is real. If, then we have

$$x + i \times y = u + i \times v$$

for some real x, y, u and v, then assuming associativity and commutativity of addition, and distributivity, we get

$$x + (-u) = i \times (v + (-y)),$$

and this forces both sides to be zero, so $x = u$ and $y = v$.

E: Hints and Solutions for Selected Exercises

1.3.6.1

Each set-theoretic identity follows from a corresponding identity of the propositional calculus, in a similar way to the first one. For instance,
$$A \cup B = B \cup A$$
follows from
$$p \vee q \Leftrightarrow q \vee p.$$
Similarly,
$$A \cup \emptyset = A$$
follows from
$$p \vee \text{false} \Leftrightarrow p,$$
and
$$A \sim (A \sim B) = A \cap B$$
follows from
$$p \wedge \neg(q \wedge \neg r) \Leftrightarrow p \wedge r.$$

Each of the propositional identities may be proven by making a truth table. The latter may, alternatively, be deduced from more basic identities by boolean algebra.

1.5.1.1

With any problem about real things, there is usually scope for interpreting the precise meaning of terms and the facts of the case. The way I look at things, only Example 3, the SU-price relation, is a function.

1.5.4.2

Suppose the class $A := \mathcal{U} \times \mathcal{U}$ of all ordered pairs were a set. The function
$$f := \{((a,b), a) : a, b \in \mathcal{U}\}$$
has A for domain and has image $\mathcal{U}$, so by the Axiom of Substitution, $\mathcal{U}$ would also be a set, in contradiction to Proposition 6.

1.5.4.3

The function defined on E by $(a,b) \mapsto a$ has image $\mathcal{S}$, so if E were a set, then so would $\mathcal{S}$ be, in contradiction to Proposition 6.

1.5.6.2

Take distinct sets a and b, and define $A_1 := \{a\}$, $A_2 := \{b\}$, and $f := \{(a,a), (b,a)\}$. Then
$$f(A_1 \cap A_2) = f(\emptyset) = \emptyset \neq \{a\} = f(A_1) \cap f(A_2).$$

1.5.6.5

Take distinct sets a and b, and then define $A := \{a, b\}$, $f := \{(a, a), (b, a)\}$, and $g := \{(a, b), (b, b)\}$.

1.7.5.2

$\{\mathcal{U}\} = \emptyset$.

2.2.2.1

Hint: This (and the following two) can be proved just by using the fact that $\leq$ is a total order on $\mathbb{N}$.

2.3.4.1

For Corollary 29, define $g_1(1, \emptyset) := a$ and

$$g_1(s(n), h) := g(s(n), h(n)), \; \forall n \in \mathbb{N} \text{ and } \forall h \in A^{(s(n)^-)},$$

and apply Theorem 28 with g replaced by g_1.

For Corollary 30, define $g_1(1, \emptyset) := a$ and

$$g_1(s(n), h) := g(h(n)), \; \forall n \in \mathbb{N} \text{ and } \forall h \in A^{(s(n)^-)},$$

and apply the Theorem with g replaced by g_1.

For Corollary 31 define

$$g_1(n, h) := g(h), \; \forall n \in \mathbb{N} \text{ and } \forall h \in A^{(n^-)},$$

and apply the Theorem with g replaced by g_1.

2.4.2.1

We have to show that $3 + 2 = 5$. Refer to Definitions 2.2 and 2.7. By Definition 2.2, $5 = s(4)$ and $4 = s(3)$. From the definition of $+$, $3 + 1 = s(3) = 4$ and $4 + 1 = s(4) = 5$, so $5 = (3 + 1) + 1$. Since $+$ is associative (by Proposition 33), this equals $3 + (1 + 1)$, and by definition $1 + 1 = s(1) = 2$, so $5 = 3 + 2$.

2.4.3.1

Hint: Use induction.

2.5.3.2

The Axiom of Substitution.

2.5.3.3

Hint: Define a suitable sequence, and use the Axiom of Substitution.

2.6.2.1

Hint: Let B be the set of boys at the dance, and G be the set of girls. For each girl s, let

$$B(s) = \{x \in B : x \text{ dances with } s\}.$$

Show that if the desired conclusion fails, then the family of the $B(s)$'s is totally-ordered by inclusion.

2.6.3.1

$\phi_4(2,5) = 2^{2^{16}} = 2^{65536}$.
$\phi_5(2,3) = \phi_4(2,4) = 2^{16} = 65536$.
$\phi_5(2,4) = \phi_4(2,65536)$, a very tall tower of powers of 2, and $\phi_5(2,5)$ is a tower of *that* many powers of 2.

3.1.5.1

1. $((a + (b \times c) + d) < (a - f)) \Rightarrow (((f \times g) = 0) \wedge (h < 0))$.
2. $((a \in (A \cup B)) \vee (b \in (C \cap D))) \Leftrightarrow (c \in (D \cap (\bigcup_n (E_n \cup F_n))))$.

3.2.1.1

Things go wrong with 0^y.

3.2.1.2

If $x + y = x$, then you can justify: $y = 0 + y = ((-x) + x) + y = (-x) + (x + y) = (-x) + x = 0$.

4.2.1.1

Induction.

4.4.1.2

Hint: Write $r = a_1 + r_1$, with $a_1 \in \mathbb{Z}_+$ and $0 \leq r_1 < 1$. This expression is unique. Now continue inductively to set

$$\frac{1}{r_n} = a_{n+1} + r_{n+1}$$

with $a_{n+1} \in \mathbb{Z}_+$ and $0 \leq r_{n+1} < 1$, as long as $r_n \neq 0$. The result of the previous exercise is useful when proving that this process must terminate.

4.4.1.3

Start as with the continued fraction, writing $r = a_1 + r_1$, and then continue inductively, setting

$$(n+1)! \cdot r_n = a_{n+1} + r_{n+1},$$

with $a_{n+1} \in \mathbb{Z}_+$ and $0 \leq r_{n+1} < 1$, as long as $r_n \neq 0$. The process must terminate after no more than denom(r) steps.

5.2.9.2

(a) Neither. (b) Below. (c) Both. (d) Below.

5.2.9.3

(a) $+\infty$, $-\infty$. (b) $+\infty$, 1. (c) 1, $\frac{1}{2}$. (d) $+\infty$, 0.

5.2.10.3

Last part: Rational x have two preimages and irrational x have one.

5.4.2.1

Hint: 1: nothing; 2 and 3 are functions.

5.4.2.2

50.

6.1.2.1

Hint: Use $0 + 0 = 0$ and cancellation.

6.1.2.2

(i) $-\frac{5}{13} + \frac{12}{13}i$. (ii) $\frac{12}{13} + \frac{5}{13}i$. (iii) $-i$. (iv) $-2 + 2i$.

6.1.2.4

$\{x + iy : x, y \in \mathbb{R} \text{ and } x^2 + y^2 = 1\}$.

D.4.1.1

Hint: $2 \times \frac{1}{2} \times \frac{1}{2}$.

D.4.3.2

Pick $n \in \mathbb{N}$ with $x \times n \in \mathbb{Z}$. Then $(x \times 1) \times n = x \times (1 \times n) = x \times n$. Now cancel n.

D.6.6.1

By (11), we may write $z = x + i \times y$ for some $x, y \in \mathbb{R}$. Then by (9s) $z \times 1 = (x + i \times y) \times 1 = x \times 1 + (i \times y) \times 1 = x + i \times (y \times 1) = x + i \times y = z$.

F: For the Teacher

This appendix is addressed not to the student but to the teacher or scholar and is about the reason I wrote this book and what it aims to do.

When I was growing up and learning mathematics it became clear that there was a rigour problem. The problem was the absence of rigour. Once you get the basic idea that everything has to be fully justified by a logically-coherent argument starting from absolutely sure principles, you start to see holes and gaps all over the place. At first, it seemed to me that the problem was confined to applied work. I watched physicists and chemists present nonsensical arguments with equanimity, because one did not expect any better of them. Sure, all matrices are diagonalizable and all Fourier series converge, if you say so. Those foolish mathematicians are wasting time trying to prove things we already know.

My first rigorous analysis course followed Olmsted's *Real Variables*[8], a rather satisfying book that began as many calculus texts do, with a set of axioms for the reals as a complete ordered field. That left the issue of the consistency of these axioms, and I was glad to find Landau's little book *Foundations of Analysis*[6], written in his famous telegraphic style, which builds up the real number system and establishes its consistency, assuming only five Peano axioms for the natural numbers. Landau took set theory for granted, and clearly set theory is 'more fundamental and basic' than numbers. I felt it important that the Peano system should be based on set axioms, or even just on logic axioms, but there were two kinds of problems.

When I was sixteen I got a prize[1] for an essay about European unity[2], and it turned out to be a book by Kasner and

---
[1] It was the 41st and last prize.
[2] I was against it, at the time.

Newman[3] about cardinality, among other things. There I learned about the contradictions inherent in naive set theory. Eventually I learned about ways to finesse these. The most satisfying for me was the account in the appendix to Kelley's *Topology* of the Kelley-Morse variant of the Hilbert-Bernays-Von Neumann idea.

The second more basic problem was Gödel's results, and the work of Tarski and Cohen. We have to accept that we cannot prove the absolute consistency or completeness of any interesting mathematical system.

OK. This means that one cannot criticise teachers who decide that they are just going to begin real analysis with the axioms of a complete ordered field. One cannot say: if you start a little further back, then the ground will be more sure.

But you can say: the Peano axioms are simpler, and provide a more satisfying starting point. This was Landau's motivation, when he decided to write his little book. He advised the student to read his account in two days, and then forget about it and get on with analysis.

My teacher Federer taught me that there is no point in writing a book if there is already a perfect account in print. He made this remark *a propos* dimension theory. He would have included an account in his own book, but he felt that Hurewitz and Wallman [2] had done it perfectly.

So why do I feel it worth writing this account? There are three reasons:

1. It is no longer acceptable to say nothing about set theory. The student must be equipped with some kind of foundation which at least avoids the easy contradictions.

I've kept the set theory to the minimum required for beginning calculus. In particular, there is no discussion of cardinality or choice.

2. I dislike the practice employed by Landau and others of progressively changing the meaning of symbols and concepts such as 2 as the development progresses. There are two things: the *consistency* of the theory, and what the numbers *are*. I think people should be free to think of 2 as a simple thing,

and to think of it as the same thing whether it be regarded as a natural number, an integer, a rational, a real or a complex number. A solution to this is to use set theory with atoms. To prove relative consistency, one constructs a model. But this should not constrain how we think of numbers. We are not obliged to regard the numbers as the same thing as the model.

3. Landau, in his 'Preface for the Scholar' (as opposed to his 'Preface for the Student') tells us that he found a gap in the inductive definition of $x + y$ that he had used for years in his lecture notes, and with the help of Von Neumann had found a way to circumvent it. He was not too happy with it, because it was a little sophisticated, and he just wanted to use 'school mathematics'. Then, at the last minute, as he was preparing his book for publication, he found a simpler way to bridge the gap, and put that in instead. Now, in fact, there is still a little gap in the proof of his Proposition 4: he implicitly uses the fact that $1 + y'$ is the same as $y' + 1$. This fact could easily be proved, using induction, at the point where it appears, so there is no real problem, and I only call it a gap because otherwise he provides full proofs at all stages. Anyway, I think I know what Von Neumann must have shown him. It is an argument that uses sets (and hence was not 'school maths' in Landau's day). It involves discussing the order on $\mathbb{N}$ before addition, whereas Landau uses addition to define the order. I would not claim that the approach I use is easier than Landau's. It is just more direct. I prefer to do the obvious thing, if possible.

It remains the case that the contents of this book are there so that the curious youngster can read through them quickly, and then forget about them and get on with analysis.

Bibliography

[1] R. Dedekind. Essays on the Theory of Numbers. Dover. 1963.

[2] W. Hurewitz and H. Wallman. Dimension Theory. Princeton University Press. 1948.

[3] E. Kasner and J.R. Newman. Mathematics and the Imagination. G. Bell and sons. 1949.

[4] J.L. Kelley. General Topology. Van Nostrand. 1955.

[5] A. Kenny. Frege. An Introduction to the Founder of Modern Analytic Philosophy. Penguin. 1995.

[6] E. Landau. Foundations of Analysis. A Supplement to Text-books on the Differential and Integral Calculus. 3rd ed. Chelsea. 1966. (translated by F. Steinhardt from the german original, Grundlagen der Analysis, 1929)

[7] J.R. Newman. The World of Mathematics I–IV. Simon and Schuster. New York. 1956.

[8] J. Olmsted. Real Variables. Appleton-Century-Crofts. 2nd edition. 1959.

Index

absolute value, 79
Ackermann, 50
Aczel, 107
addition, 42, 53, 63, 79
appendix, 98, 101, 105, 109, 118, 125
Archimedes, 85, 95
arithmetic, 90
associative, 42, 46, 58, 99
atom, 5, 57, 71, 89, 106
atom with structure, 107
axiom, amalgamation, 12
axiom, classification, 8
axiom, completeness, 75
axiom, complex arithmetic, 90
axiom, complex atoms, 89
axiom, comprehension, 8
axiom, dense rationals, 74
axiom, extent, 7
axiom, foundation, 1
axiom, induction, 31
axiom, integers, 52
axiom, membership, 7
axiom, natural numbers, 30
axiom, order, 71
axiom, ordered pairs, 16
axiom, power set, 12

axiom, rationals, 58
axiom, real atoms, 71
axiom, substitution, 19
axiom, successor, 31
axiom, trichotomy, 52
axiom, union, 12
axioms, 101, 105, 109

Barwise, 2
Bechtluft-Sachs, 51
belongs, 5
Bernays, 126
bijection, 21
binary operation, 41, 58
binomial coefficient, 67
binomial theorem, 93
BODMAS, 54
Boole, 30
bound, lower, 27, 74
bound, upper, 27, 74
bounded, 74

cancellation, 43, 46, 60, 84
Cantor, 1
Cartesian product, 16, 73
Cauchy sequence, 108
ceiling, 86
characteristic, χ_A, 57

class, 6
coefficient, 92
Cohen, 126
commutative, 42, 46, 58, 99
completeness, 75
complex conjugate, 91
complex number, 89, 108
composition, 21
compound proposition, 99
conjugate, complex, 91
conjunction, 98
conjunction,∧, 2
consistency, 105, 126
continued fraction, 69
continuum hypothesis, 1
contradiction, 99

De Morgan, 13, 99
Dedekind, 108
Dedekind cut, 77
definition, 105
definition, inductive, 39
denominator, 61
dense, 74
Descartes, 73
diagonal, 21
difference, 9
disjoint union, 13
disjunction, 98
disjunction,∨, 2
Diskworld, 96
distributive, 46, 99
division, 84
domain, 18
doubleton, 13

eight, 32
element, 5

empty class,∅, 8
equality,=, 3
equivalence relation, 24
Etchemendy, 2
excluded middle, 99
existential quantifier,∃, 4
exponent, 47, 66, 91
extended complex, 95
extended real numbers, 88

factorial, 46, 55
Federer, 126
field, 71
final set, 33
finite, 32
five, 32
floor, 86
foundation, axiom of, 1
four, 32
fraction, 59, 62
fraction, continued, 69
Frege, 30
function, 18
function, Ackermann, 50
function, characteristic, 57
function, inverse, 22
function, rational, 92

Gödel, 126
Gauss, 89
greatest element, 28
greatest lower bound, 74

Hamilton, 89
Hilbert, 126
Hurwitz, 126

identity, 58
identity function, 21

INDEX

image, 18
imaginary, 91
induction, 31, 34, 44
inductive definition, 39
inequality, triangle, 81
infimum, 74, 77
infinite, 32
infinity, 88, 95, 109
infinity, ∞, 72
infix, 41
initial set, 33
initial value, 39
injection, 21
injective, 21
integer, 52, 108
intersection, 9
interval, 72
inverse, 59
inverse function, 22

Kasner, 126
Kelley, 126
Kronecker, 70

Landau, 125, 126
laws of exponents, 47, 92
least element, 28
least upper bound, 74
liar paradox, 2
logic, 2
logical rule, 105
lower bound, 27, 74
lowest terms, 62

Maple, 28
Marx, 8
mathematical theory, 105
maximum, 28, 49

member, 5
minimum, 28, 49
minus, 53, 62, 78
model, 106, 109
multiplication, 45, 54, 58, 81

naive set theory, 126
natural number, 30
negation, 98
negation, $\neg$, 2
Newman, 126
nine, 32
NKMPC, 6
nonnegative, 53
nonpositive, 53
number, complex, 89
number, natural, 30
number, rational, 58
number, real, 70
numerator, 61

Olmsted, 125
one-to-one, 21
onto, 21
operation, binary, 41, 58
operation, unary, 41
order, 37, 53, 65, 71
order, partial, 26
order, strict, 27
order, total, 27
ordered field, 71
ordered pair, 16, 106

partial order, 26
Pascal's triangle, 68
Peano, 30, 125
Perelman, 2
polynomial, 92

positive, 53
power, 47, 55, 66, 85, 91
power set, 2^A, 12
Pratchett, 96
precedence, 54
predicate, 4
product, 45
proper class, 5, 14
proposition, 2, 98
proposition, compound, 99
propositional calculus, 98

range, 18
rational, 67
rational function, 92
rational number, 58, 108
real number, 108
real numbers, 70, 87
reciprocal, 59
redundancy, 109
relation, 16, 86
relation, equivalence, 24
Russell's paradox, 14

sequence, 48, 92
sequence, Ackermann, 50
set, 1, 7
set theory, 126
seven, 32
singleton, 9
six, 32
strict order, 27
subclass, 7
subtraction, 44, 54, 64, 81
successor, 31
supremum, 74, 77

Tarski, 50, 126

teacher, 125
ten, 32
theorem, 105
theorem, binomial, 93
theory, mathematical, 105
thing, 1
three, 32
total order, 27
triangle inequality, 81
trichotomy, 52
truth table, 100
truth value, 98
two, 32, 126

UCC, 30
unary operation, 41
unbounded, 74
undefined term, 105
union, 9
universal class,$\mathcal{U}$, 8
universal quantifier,$\forall$, 4
upper bound, 27, 74

Vandermonde, 95
Venn diagram, 10
Von Neumann, 126

Wallman, 126

Anthony G. O'Farrell was born in Dublin in 1947, and was educated by the Sacred Heart Sisters in Roscrea, the Irish Sisters of Charity in Templemore, and the Irish Christian Brothers in Templemore and Drimnagh Castle. He worked in the Irish Meteorological Service from 1964 to 1968. For most of that time he was a student of Mathematical Science at University College, Dublin, where he was particularly influenced by P. G. Gormley, who turned him decisively to analysis, particularly complex analysis. Resigning from the Meteorological Service to devote himself to Mathematics, he was further influenced by T. J. Laffey and E. C. Schlesinger, and decided to continue his studies in the United States. At Brown University, his horizons were expanded by H. Federer, B. Harris, A. Browder, J. Wermer, B. Cole, W. Fulton and A. Landman, among others, and his subsequent work focussed mainly on algebraic and geometric aspects of real and complex analysis. In 1975, after two years at the University of California at Los Angeles, he was appointed Professor of Mathematics at Maynooth College, He was elected to the Royal Irish Academy in 1980. While engaged in teaching, research and administration at Maynooth, he contributed to the mathematical and scientific community in Ireland and abroad, and visited research institutes and universities in Canada, France, Germany, India, Israel, Japan, Russia, Sweden, Spain, the UK and USA, as well as speaking at many international conferences and collaborating with many researchers.

Now Professor Emeritus of Maynooth University, he devotes himself to research, writing and *pro-bono* activities, as well as a modicum of outdoor activity and cultural pursuits. He is married, with three surviving children and seven grandchildren.

www.ingramcontent.com/pod-product-compliance
Lightning Source LLC
Chambersburg PA
CBHW060858170526
45158CB00001B/408